纪念吴文俊先生诞辰100周年

国家出版基金项目
NATIONAL PUBLICATION FOUNDATION

The Complete Works of Wu Wen-Tsun
Game Theory, Algebraic Geometry, and Planar Imbedding of Graphs

吴文俊全集·博弈论、代数几何、图的平面嵌入卷

吴文俊 著

李文林 高小山 编订

科学出版社
龙门书局

北 京

内 容 简 介

本书收录吴文俊发表的博弈论、代数几何和图的平面嵌入等三个方面的论文. 其中,《关于博弈理论基本定理的一个注记》是中国博弈论研究的开山之作.《活动受限制下的非协作对策》等两篇论文则包含了吴文俊对博弈论最重要的贡献——本质均衡概念的提出及其存在性证明. 本质均衡是一类特殊的纳什均衡, 吴文俊是能在早期就认识纳什理论的深刻意义并率先作出有生命力的贡献的少数数学家.《具有对偶有理分割的代数簇》等一组代数几何学论文, 记载了吴文俊在代数几何领域的成果, 其中最重要的就是对于含奇点的代数簇定义了陈省身示性类. 本书最后的《集成电路设计中的一个数学问题》《线性图的平面嵌入》等4篇文章, 是将拓扑嵌入理论应用于集成电路布线问题的成果.

本书可作为数学工作者、数理经济学者和相关工程技术人员的参考用书, 也适合大学教师、研究生和广大科学爱好者阅读.

图书在版编目（CIP）数据

吴文俊全集·博弈论、代数几何、图的平面嵌入卷/吴文俊著；李文林, 高小山编订. —北京：龙门书局, 2019.5
ISBN 978-7-5088-5560-8

Ⅰ. ①吴⋯ Ⅱ. ①吴⋯ ②李⋯ ③高⋯ Ⅲ. ①博弈论-文集 ②代数几何-文集 Ⅳ. ①O1-53

中国版本图书馆 CIP 数据核字(2019) 第 074603 号

责任编辑：李 欣 赵彦超／责任校对：邹慧卿
责任印制：赵 博／封面设计：无极书装

科学出版社 出版
龙门书局
北京东黄城根北街 16 号
邮政编码：100717
http://www.sciencep.com
北京建宏印刷有限公司印刷
科学出版社发行　各地新华书店经销
*

2019 年 5 月第 一 版　开本：720×1000 1/16
2025 年 1 月第三次印刷　印张：9 1/4
字数：190 000
定价：88.00 元
(如有印装质量问题, 我社负责调换)

编 者 序

中国现代数学的崛起,开始于20世纪初,经历了几代人坚苦卓绝的努力. 在这百年奋战中涌现出来的数学家中,吴文俊是最杰出的代表之一. 他早年留学法国,留学期间就已在拓扑学方面做出了杰出贡献,提出了后来以他的名字命名的"吴公式"和"吴示性类". 回国后提出了"吴示嵌类"等拓扑不变量,发展了统一的嵌入理论. 他关于示性类与示嵌类的研究,已成为 20 世纪拓扑学的经典,至今还在前沿研究中使用. 20世纪70年代以来,吴文俊院士在汲取中国古代数学精髓的基础上,开创了崭新的现代数学领域——数学机械化. 他发明的被国际上誉为"吴方法"的数学机械化方法,改变了国际自动推理的面貌,形成了自动推理的中国学派,已使中国在数学机械化领域处于国际领先地位. 上述工作无疑属于 20 世纪中国数学赶超国际先进水平的标志性成果,而吴文俊院士博大精深的科学研究,除了拓扑学与数学机械化以外,还跨越了代数几何、博弈论、中国数学史、计算图论、人工智能等众多领域,并在每个领域都留下了这位多能数学家的重要贡献.

吴文俊先生是一位具有强烈爱国精神的数学家. 自 1950 年谢绝法国师友的挽留回到祖国后,半个世纪如一日,为在他深爱的中华故土发展数学事业而鞠躬尽瘁. 除了第一流的科研成果,吴文俊先生长期身处中国数学界领导地位,在团结带领整个中国数学界赶超世界先进水平方面,也做出了不可磨灭的贡献. 特别是,吴文俊先生在担任中国数学会理事长期间,领导中国数学最终成功地加入了国际数学联盟,此举大大提高了我国数学界的国际地位,同时也为我国成功举办 2002 年国际数学家大会铺平了道路.

吴文俊治学严谨,学术思想活跃,无论获得多么高的声誉,他总是勤奋地在科研第一线工作,一生积极进取,锲而不舍,不断取得新的成就. 在开始从事机器证明时,他已近花甲之年,从零开始学习编写计算机程序,每天十多个小时在机房连续工作,终于在几何定理机器证明这一难题上取得成功.

吴文俊先生为中国现代数学的发展建立了丰功伟绩,而他本人却始终淡泊、谦逊. 他处事公正豁达,待人充满善意,受过他帮助的人可以说不计其数. 正因如此,这位有着崇高国际声望而平易近人的学者,受到了每一个认识他的人格外的爱戴与尊敬.

2019 年 5 月 12 日是吴文俊先生百年诞辰. 为了纪念这个特殊的日子,我们编辑出版了《吴文俊全集》,通过系统地收录、整理吴文俊先生的学术著作和论文,纪念吴先生的学术思想及学术成就. 全集共计 13 卷, 包括拓扑学 4 卷、数学机械化 5 卷以及数学史、博弈论与代数几何、数学思想各 1 卷; 同时, 全集还设有附卷, 收录吴文俊先生的同事、学生和其他社会各界人士发表过的与吴先生有关的各类文献资料.

最后, 我们对在全集编辑中给予帮助的各位同事表示衷心感谢; 感谢国家出版基金对于全集出版的资助; 感谢科学出版社编辑人员在出版全集时认真细致的专业精神; 感谢相关出版与新闻机构在版权方面提供的帮助.

李邦河　高小山　李文林

2019 年 3 月

前　言

　　吴文俊先生属于渊博多能的数学家之列，除了拓扑学、数学机械化和数学史，他的科学贡献还涉及博弈论、代数几何、图论等多个领域. 本卷收录吴文俊发表的博弈论论文 3 篇、代数几何论文 4 篇、图的平面嵌入论文 4 篇.

　　吴文俊的第一篇博弈论论文《关于博弈理论基本定理的一个注记》发表于 1959 年，当时国际上方兴未艾的博弈论在国内尚无人问津，吴文俊的这篇论文遂成中国博弈论研究的开山之作. 本卷收录的《关于博弈理论基本定理的一个注记》和 *Essential Equilibrium Points of n-Person Non-cooperative Games* 两篇论文则包含了吴文俊对博弈论最重要的贡献——本质均衡概念的提出及其存在性证明. 本质均衡是一类特殊的纳什均衡，吴文俊 (与江嘉禾合作) 的上述工作事实上是所谓纳什均衡精练研究方面的最早结果，到 20 世纪 80 年代又重新受到国际学界的关注，至今仍为世界一流博弈论学者引用不绝. 今天，纳什几乎是家喻户晓的人物，但在 20 世纪五六十年代之交，他的理论可以说还是曲高和寡，吴文俊是能在早期就认识纳什理论的深刻意义并率先做出有生命力的贡献的少数数学家.

　　本卷收录的《具有对偶有理分割的代数簇》等一组代数几何学论文，记载了吴文俊在代数几何领域的成果，其中最重要的就是对于含有奇点的代数簇定义了陈省身示性类. 以多项式组的公共零点集 (代数簇) 为主要对象的代数几何，由于拓扑学方法的引入而成为国际上发展迅猛的现代数学热门领域，但在吴文俊介入这一领域的时候，代数几何研究在国内几乎是一片空白. 吴文俊可以说是 "单刀赴会"，独自一人从学习国外文献开始，凭借着自己的拓扑学功底，愣是在短短几年内啃下了"硬骨头". 吴文俊关于代数几何的研究在后来的数学机械化理论的建立过程中起了重要作用并有进一步发展.

　　本卷最后的《集成电路设计中的一个数学问题》《线性图的平面嵌入》等 4 篇文章，是将拓扑嵌入理论应用于集成电路布线问题的成果. 集成电路布线问题实际是一个线性图的平面嵌入问题. 吴文俊运用示嵌类理论把问题归结为简单的模 2 方程的计算问题，不仅可得出是否可嵌入的判据，而且可以指出如何具体地布线. 他的方法完全可以计算，可以上计算机.

　　值得指出的是，吴文俊在上述三个领域的研究成果都是在特殊的背景下进行和完成的. 博弈论的研究完成于 1958 年开始的 "大跃进" 时期，在提倡理论联系实际的形势下，他选择了运筹学. 运筹学是应用性很强的学科，不过当时国内主要关注的是线性规划，而吴文俊却将目光投向了拓扑学有用武之地的博弈论并迅速锁定于风乍起的纳什理论; 代数几何研究的缘起是吴先生在中国科技大学开设代数几何课程, 按照他本人的说法是 "边学习边研究" 的硕果; 至于平面布线问题的研究, 则是他在理论研究陷于停滞的 "文革" 期间进行的. 在一次 "大批判" 会议中, 坐在会议室一隅的吴文俊随手翻阅书架上的一本刊物, 其中有一篇介绍印刷电路的文章提到集成电路中成千上万的元件用线连成网络, 这些线路是否能分布在一个平面上的问题, 他立刻意识到这正是自己熟悉的平面嵌入问题, 于是开展研究. 由于当时学术期刊停刊, 其研究结果到 1973 年才发表.

所有这些都反映了吴文俊先生豁达睿智的个性和敏锐深邃的学术洞察力, 同时也说明了他对理论联系实际的正确认识与恰当把握, 这使他在任何情况下都不会消极懈怠, 而是积极进取, 去寻找合适的课题并能迅速抓住本质, 占领前沿. 吴文俊在上述三个领域的研究成果虽然只占整个科学贡献的一小部分, 但却在每个领域留下了历史的脚印, 也使我们看到了一个真正的学者的风格.

李文林　高小山

2019 年 4 月 12 日

目 录

关于博弈理论基本定理的一个注记 ·· 1

活动受限制下的非协作对策 ·· 5

Essential Equilibrium Points of n-Person Non-cooperative Games ·········· 26

具有对偶有理分割的代数簇 ·· 41

代数簇上的陈省身示性系 ·· 50

Chern Classes on Algebraic Varieties with Arbitrary Singularities ·········· 59

On Chern Numbers of Algebraic Varieties with Arbitrary Singularities ······ 61

集成电路设计中的一个数学问题 ·· 75

线性图的平面嵌入 ·· 102

On the Planar Imbedding of Linear Graphs ································ 106

On the Planar Imbedding of Linear Graphs (Continued) ···················· 122

关于博弈理论基本定理的一个注记*

§1. von Neumann 在其博弈理论 (theory of games) 中对极大极小基本定理 [1,2] 曾经有各种方式的推广，但都需要关于策略空间或胜负函数或两者很强的代数的假设，例如某种线性或凸性假设之类 (参阅 [3—14])．本文提供了另一种推广，其性质是纯粹拓扑的，与所有已知的推广不同，特别与纯粹代数性质的 H.Weyl 的推广相反．我们的证明只用到了点集的初等性质，而并没有用定点定理或类似的定理，也没有用到关于凸集的定理．

§2. 设 R 是一个闭线段．我们知道 R 可拓扑地表示为只有两个非断点 (cut point) 的绵续统，此两点即为 R 的端点．在 R 上于是恰有两种方式来引入一个次序关系，使对 R 的任一子集 E，对于任一 R 上选定的次序关系 $<$ 与 $>$，下确界 $\inf E$ 与上确界 $\sup E$ 都有明确意义．在以下所选次序将固定不变，对此由 $z \geqslant \lambda, z > \lambda, z < \lambda'$ 与 $z \leqslant \lambda'$ 所定的 R 的子集将各记为 $\bar{R}_\lambda^+, R_\lambda^+, R_\lambda^-$ 与 $\bar{R}_\lambda^-$．对任意空间 X, Y 与任意 $X \times Y$ 到 R 中的映象 f，不论连续与否，将用记号 $f(x,y) = f_x(y) = f_y(x), x \in X, y \in Y$．我们将称 f 对 X 是强连通的，如果它具有以下二性质：

(P_1) 对任意 $a, b \in X$，有一以 $\bar{a}, \bar{b}$ 为端点的闭线段 I 到 X 中的连续映象 h，使 $h(\bar{a}) = a, h(\bar{b}) = b$ 且对任意 $y \in Y$ 与任意 $\lambda \in R, h^{-1} f_y^{-1}(\bar{R}_\lambda^+)$ 在非空集时，必为一连通集 (这隐含了 X 是弧连通的)．

(P_2) 对任意有限个点 $x_1, \cdots, x_k \in X$ 与任意 $\lambda \leqslant R, f_{x_1}^{-1}(R_\lambda^-) \cap \cdots \cap f_{x_k}^{-1}(R_\lambda^-)$ 在非空集时，为一连通集．

于是我们所谓 von Neumann 定理的推广可述如下：

定理 设 Y 为一紧致可分空间而 X 为弧连通的．若 $f : X \times Y \to R$ 对 X 强连通，而 f_x, f_y 对任意 $x \in X, y \in Y$ 皆连续，则

$$\operatorname{Inf}\{\operatorname{Sup} f_y(X)/y \in Y\} = \operatorname{Sup}\{\operatorname{Inf} f_x(Y)/x \in X\}. \tag{1}$$

§3. 定理的证明有赖于下述二引理．

引理 1 设 X 为一以 a, b 为端点的闭线段而 Y 为一可分空间，又设 λ 为 R 的一个定点而 f 为 $X \times Y$ 到 R 中的一个映象，使对任意 $x \in X$ 与 $y \in Y, f_y^{-1}(\bar{R}_\lambda^+)$ 与 $f_x^{-1}(R_\lambda^-)$ 在非空时皆为连通集，f_x 与 f_y 皆连续，且 $f_y^{-1}(\bar{R}_\lambda^+)$ 不含 a 即含 b．于是必有一点 $\xi \in X$，使 $f_\xi(Y) \subset \bar{R}_\lambda^+$．

---

*本文原载《科学记录》，新辑第 3 卷，第 6 期，1959 年．

证. 设 $y_1, y_2, \cdots, y_n, \cdots$ 为 Y 中一个可数稠密集. 置 $I_i = f_{y_i}^{-1}(\bar{R}_\lambda^+)$, 则 I_i 连通且不含 a 即含 b, 因之为一含 a 或 b 的一个线段或即 a 或 b 自身. 取 n 固定并设 $I_1 \cap \cdots \cap I_n$ 为空集, 则必有集 $\{1, 2, \cdots, n\}$ 中的某整数 α, β 使 $I_\alpha \cap I_\beta = \varnothing$, 因之有 $x_n \in X$ 使 $f_{y_\alpha}(x_n) \in R_\lambda^-, f_{y_\beta}(x_n) \in R_\lambda^-$. 故 $J = f_{x_n}^{-1}(\bar{R}_\lambda^-)$ 含有 y_α, y_β 二者且由假设连通. 设 $a \in I_a$, 则 $a \notin I_\beta, b \notin I_\beta, b \notin I_\alpha$ 而 $J_\alpha = J \cap f_\alpha^{-1}(\bar{R}_\lambda^+)$ 含 $y_\alpha, J_\beta = J \cap f_b^{-1}(\bar{R}_\lambda^+)$ 含 y_β. 由假设 J_α, J_β 为 J 中闭集且 $J_\alpha \cup J_\beta = J$. 由此得知 $J_\alpha \cap J_\beta \neq \varnothing$. 取 $\eta \in J_\alpha \cap J_\beta = J$. 则 $I = f_\eta^{-1}(\bar{R}_\lambda^+)$ 含有 a 与 b 二者, 因 I 连通, 故 I 与 X 相合. 由此得 $x_n \in I = f_\eta^{-1}(\bar{R}_\lambda^+)$ 或 $\eta \in f_{x_n}^{-1}(\bar{R}_\lambda^+)$, 与 $\eta \in J = f_{x_n}^{-1}(\bar{R}_\lambda^-)$ 相悖. 因之 $I_1 \cap \cdots \cap I_\eta \neq \varnothing$ 而可取 $\xi_n \in I_1 \cap \cdots \cap I_n$ 使 $f_{y_i}(\xi_n) \in \bar{R}_\lambda^+, i = 1, 2, \cdots, n$. 命 ξ 为 $\xi_n, n = 1, 2, \cdots$ 的一个极限点, 则对一切 i 应有 $f_{y_i}(\xi) = f_\xi(y_i) \in \bar{R}_\lambda^+$. 因集 $\{y_i\}$ 在 Y 中稠密, 故对一切 $y \in Y$ 有 $f_\xi(y) \in \bar{R}_\lambda^+$, 如所欲证.

引理 2_n 设 Y 为一可分空间而 X 弧连通, 又设 λ 为 R 的一个定点而 $f: X \times Y \longrightarrow R$ 为一对 X 强连通的映象. 若 f_x, f_y 对一切 $x \in X, y \in Y$ 皆连续, 且有 n 个点 $a_1, \cdots, a_n \in X$ 使 $f_{a_1}^{-1}(\bar{R}_\lambda^+) \cup \cdots \cup f_{a_n}^{-1}(\bar{R}_\lambda^+) = Y$, 则有一点 $\xi \in X$ 使 $f_\xi(Y) \subset \bar{R}_\lambda^+$.

证. 我们将对 n 施行归纳法. 引理在 $n = 1$ 时是平凡的. 假设引理在 $n - 1$ 时为真, 此处 $n \geqslant 2$. 置 $Y' = f_{a_n}^{-1}(\bar{R}_\lambda^-)$, 则引理 2_{n-1} 中的假设对空间 X, Y', 点 $\lambda \in R$, 映象 $f/X \times Y'$, 与点 $a_1, \cdots, a_{n-1} \in X$ 而言成立, 故有一点 $\xi' \in X$ 使 $f_{\xi'}(Y') \subset \bar{R}_\lambda^+$. 由映象 f 的强连通性的 (P_1), 有一以 a, b 为端点的闭线段 I 到 X 中的连续映象 h 使 $h(a) = a_n, h(b) = \xi'$, 而 $(f_y h)^{-1}(\bar{R}_\lambda^+)$ 对任意 $y \in Y$ 在非空集时即为连通集. 由此对空间 (I, Y), 点 $\lambda \in R$ 与映象 $\bar{f}: I \times Y \longrightarrow R$, 此处 $\bar{f}(x, y) = f_y h(x), x \in I, y \in Y$, 引理 1 的假设满足, 故有一点 $\bar{\xi} \in I$ 使 $\bar{f}_{\bar{\xi}}(Y) \subset \bar{R}_\lambda^+$. 对 $\xi = h(\bar{\xi})$ 即有 $f_\xi(Y) \subset \bar{R}_\lambda^+$, 如所欲证.

§4. 至此, 我们的定理即易从通常的推理得出. 盖设 (1) 不真, 则

$$\text{Inf}\{\text{Sup} f_y(X)/y \in Y\} > \text{Sup}\{\text{Inf} f_x(Y)/x \in X\},$$

于是有 $\lambda \in R$ 使

$$\text{Inf}\{\text{Sup} f_y(X)/y \in Y\} > \lambda > \text{Sup}\{\text{Inf} f_x(Y)/x \in X\}. \tag{2}$$

对任意 $y \in Y$ 即有一点 $x_y \in X$ 使 $f(x_y, y) \in R_\lambda^+$. 因每一 $U_y = f_{x_y}^{-1}(R_\lambda^+) \subset Y$ 皆为开集, 故由 Y 的紧致性有有限个点 $y_1, \cdots, y_n \in Y$ 使 $a_i = x_{y_i}$ 时, $U_{y_i} = f_{a_i}^{-1}(R_\lambda^+), i = 1, 2, \cdots, n$, 覆盖 Y, 而 $f_{a_i}^{-1}(\bar{R}_\lambda^+)$ 更是如此. 于是对空间 X, Y, 点 $\lambda \in R$, 映象 f 与点组 $a_1, \cdots, a_n \in X$ 而言, 引理 2_n 的假设满足. 故有一点 $\xi \in X$ 使 $f_\xi(Y) \subset \bar{R}_\lambda^+$. 由此得 $\text{Sup}\{\text{Inf} f_x(Y)/x \in X\} \geqslant \lambda$ 与 (2) 相悖. 因之 (1) 成立而定理得证.

§5. 例与注.

(A) 设空间 X', Y' 与映象 $f': X' \times Y' \longrightarrow R$ 满足定理的条件. 设 X, Y 为任意各在拓扑映象 φ, ψ 下与空间 X', Y' 同胚的空间而 h 为 R 的一个保序拓扑变换. 定义 $f: X \times Y \longrightarrow R$ 如 $f(x, y) = hf'(\varphi(x), \psi(y)), x \in X, y \in Y$. 则 (X, Y, f) 亦满足定理的条件. 这说明了上述 von Neumann 定理推广的纯粹拓扑性质.

(B) 设 X, Y 为线性拓扑空间中的凸集, f 为 $X \times Y$ 上的实函数, 对 X 拟凹而对 Y 拟凸 (quasi-concave 与 quasi-convex, 见 Nikaidô, [12]), 则 f 对 X 强连通. 因之, 我们的定理包括了在 Y 可分的假设之下 Nikaidô 关于 von Neumann 定理的推广, 同样也包括了 Ville, Wald, Kneser 等人的推广. 另一方面, 因为我们的定理不含任何代数的假定而其他则否, 故与其他的推广无关.

(C) 几乎所有的推广都假定策略空间 X, Y 是凸的, 因之, 拓扑上说来可缩成一点. 下面的例则并非如此, 但能满足我们定理中的要求. 设 X 与 Y 为圆周. 试将环面 $X \times Y$ 依通常方式嵌入直角坐标 (x, y, z) 的三维欧氏空间使所得环面的轴为 x 轴而其平行圆与集 $X \times (y), y \in Y$ 相当. 于是 z 坐标定义了一个在 $X \times Y$ 上的实函数 f, 满足定理中的条件. 唯一的最好策略可见与 f 的两个鞍点之一相当. 若以 x 坐标定义实函数, 则最好策略集将与环面上的一个圆相当.

(D) 由于不加任何代数的条件, 因之对空间 X, Y 与函数 f 作若干连通性的假设以保证等式 (1) 的成立是很自然的. 但下例说明我们对连通性的条件是过强了. 设 X, Y 为圆周如前. 试将环面 $X \times Y$ 表现为一正方形 $ABCD$, 其两组对边各予恒同. 作一 $X \times Y$ 上的连续实函数 f 使在正方形 $ABCD$ 的四边及其对角线 AC 上 $f = 0$, 而在三角形 ABC 与 ACD 的内部, 各有 $f > 0$ 与 $f < 0$, 且各有一极大与一极小, 并设等值线 $f = c$ 为各边, 各与 ABC 或 ACD 的各边平行的三角形, 视 $c > 0$ 或 $c < 0$ 而定. 于是强连通中的条件 (P_2) 能满足, 但 (P_1) 则否. 但等式 (1) 则仍成立, 且其最好策略集仅含一点, 此点与正方形的四个顶点相当, 而博弈的值则为 0. 同样若将 (C) 的首例中 X 与 Y 的作用互易, 其情形亦同.

参考文献

[1] von Neumann, J. *Math. Ann*, 1928, 100: 295-320.

[2] von Neumann, J. and Morgenstern, O. *Theory of Games and Economic Behavior*, 3rd, ed., Princeton, 1953.

[3] Ville, J. *Sur la théorie générale des jeux où intervient l'habileté des joueurs, Traité du calcul des probabilités et de ses applications*, 1938, 2(5): 105-113, Paris.

[4] Kakutani, S. . *Duke Math.* J., 1941, 8: 457-459.

[5] Wald, A. *Ann. of Math.*, 1945, 46: 281-286.

[6] Karlin, S. Operator treatment of minimax principle, *Contributions to the theory of games*, 1950, Vol.1, Princeton.

[7] Weyl, H. Elementary proof of a minimax theorem due to von Neumann. *Contributions to the theory of games*, 1950, vol.1, Princeton.
[8] Kneser, H. *C. R. Paris*, 1952, 234: 2418-2420.
[9] Fan, K. *Proc. Nat. Acad. Sci.*, 1952, 38: 121-126.
[10] Fan, K. *Proc. Nat. Acad. Sci.*, 1953, 39: 42-47.
[11] Berge, C. *Bull. Soc. Math. France*, 1954, 82: 301-319.
[12] Nikaidô, H. *Pacific J.Math.*, 1954, 4: 65-72.
[13] Peck, J. E. L. and Dulmage, A.L. *Canadian J.Math.*, 1957, 9: 450-458.
[14] Sion, M. *Pacific I.Math.*, 1958, 8: 171-176.

活动受限制下的非协作对策[*]

§1. 引言

设 Γ 是一 n 人对策,第 i 人的策略空间是 S_i,赢得函数是 $H_i(x_1,\cdots,x_n), x_i \in S_i, i=1,\cdots,n$. 命 S_i^* 为第 i 人的一个混合策略集,而 $H_i^*(\mu_1,\cdots,\mu_n)$, $\mu_i \in S_i^*$,为其相应数学期望. 按 Nash[5],策略组 $\mu^* = (\mu_1^*,\cdots,\mu_n^*)$ 称为对策 $\Gamma = \langle I, \{S_i\}, \{H_i\}\rangle$(这里 $I = \{1,\cdots,n\}$ 是对策者集) 的一个平衡局势,如果对每一 $\mu_i \in S_i^*, i=1,\cdots,n$,有

$$H_i^*(\mu_1^*,\cdots,\mu_n^*) \geqslant H_i^*(\mu_1^*,\cdots,\mu_i,\cdots,\mu_n^*), \quad i=1,\cdots,n.$$

换言之,这些策略 μ_i^* 的选择使第 i 人无意改变他的策略,只要其余对策者不改变他们的策略的话. Nash 关于对策论的一条基本定理 [5] 说: 如果每一 S_i 都是有限集,而 S_i^* 是所有可能的混合策略的集合,则对策的平衡局势必然存在. Glicksberg [3] 曾将此定理推广至下述情形: S_i 都是 Hausdorff 复紧空间,S_i^* 是 S_i 上所有 Borel 集所成 σ-域上一切正则概率测度的集体,而 H_i 都是积空间 $S_1 \times \cdots \times S_n = S$ 上的连续函数.

在 Nash 与 Glicksberg 的情形中,(混合) 策略的选择与改变因之都是任意的. 但若假定策略的选择与改变都受有某种限制,则应更接近于现实. 本文目的即在讨论这种所谓活动受限制下的对策的平衡局势,这种对策的精确定义见 §8. 如本文所示,这种对策的平衡局势可以不存在,而且决定平衡局势存在与否的主要因素应该是活动限制区域间错综复杂的关系,而与策略空间本身的复杂程度无关. 特别在 Nash-Glicksberg 情形下所以能保证平衡局势的存在,正是因为限制区域十分简单的缘故,这时的限制区域事实上只有一个,即各对策者的对策空间全部,虽然这时的对策空间本身可以是任意 Hausdorff 复紧空间.

§8 中主要定理的证明依循着奠基于 Kakutani 定点定理推广的通常推理. 但这里需用到代数拓扑中远为深刻的工具,对此可参考 J. Leray 原著 [4]. 关于泛函方面,我们主要引征 [2] 一书.

---

[*] 本文原载《数学学报》,第 11 卷第 1 期,1961 年 3 月.

§2. 概率测度的支柱

设 X 是一 Hausdorff 复紧空间, 而 $B(X)$ 是 X 上一切 Borel 集所成的 σ-域. 对定义在 $B(X)$ 上的任一正则概率测度 μ, 我们将以 $[\mu]$ 表所有点 $x \in X$ 的集合, 对这些点 x 的任一邻域 U 有 $\mu(U) \neq 0$. 我们称这一集合为 μ 的支柱.

引理 X 中 $B(X)$ 上一个正则概率测度 μ 的支柱 $[\mu]$ 满足以下性质:

1) $[\mu]$ 是 X 的闭集.
2) $[\mu]$ 是 X 中使 $\mu(F) = 1$ 的一切闭集 F 的交.
3) 对包含 $[\mu]$ 的任意开集 U 有 $\mu(U) = 1$.
4) $\mu([\mu]) = 1$.
5) $\mu(X - [\mu]) = 0$.
6) $[\alpha\mu + \beta v] \subset [\mu] \cup [v]$, 这里 μ, v 是 $B(X)$ 上任两正则概率测度, α, β 是任两实数, 满足 $\alpha, \beta \geqslant 0, \alpha + \beta = 1$.

证. 由定义, 若 $x \not\in [\mu]$, 则 x 有一邻域 U_x 使 $\mu(U_x) = 0$, 于是任一 U_x 中的 x' 都 $\not\in [\mu]$. 因之 $X - [\mu]$ 是开集或 $[\mu]$ 是闭集, 而 1) 得证.

命 G 为 X 中使 $\mu(F) = 1$ 的一切闭集 F 的交集. 若 $x \not\in G$, 则有一闭集 $F \subset X$ 使 $x \not\in F$, 而 $\mu(F) = 1$. 因之对任一与 F 不相遇的 x 的邻域 U_x 有 $\mu(U_x) = 0$. 由定义有 $x \not\in [\mu]$, 故 $[\mu] \subset G$. 另一方面, 若 $x \not\in [\mu]$, 则有含 x 的开集 U_x 使 $\mu(U_x) = 0$. 因之对闭集 $F = X - U_x$ 有 $\mu(F) = 1$. 因 $x \not\in F$, 故更有 $x \not\in G$. 由此得 $[\mu] \supset G$. 故 $[\mu] = G$ 而 2) 得证.

命 U 为任一包含 $[\mu]$ 的开集. 由定义与 1) 对任意 $x \in U$ 应有含 x 的开集 U_x 使 $\mu(U_x) = 0$ 与 $U_x \cap [\mu] = \emptyset$. 这些 U_x 的全体构成 $X - U$ 的一个开覆盖. 因 X 是复紧的, 故 $X - U$ 也是复紧的, 因之有有限 $U_i = U_{x_i}, i = 1, \cdots, n$, 使 $\{U_i\}$ 已足以构成 $X - U$ 的一个开覆盖. 于是 $\mu(X - U) \leqslant \sum \mu(U_i) = 0, \mu(X - U) = 0$ 或 $\mu(U) = 1$. 这证明了 3).

今设 $\mu([\mu]) < 1$. 因 μ 是正则的, 故有含 $[\mu]$ 的开集 U 使 $\mu(U) < 1$. 这与 3) 相违而 4) 得证. 推断 5) 即由 4) 而来.

因 6) 甚显然, 故引理证毕.

§3. 支柱在一指定集合中的概率测度集

设 X 为 Hausdorff 复紧空间而 $B(X)$ 为 X 上一切 Borel 集所成的 σ-域. 对于 $B(X)$ 上任意正则可数加的有界集合函数 μ, 命 $v(\mu, X)$ 为 μ 在 X 上的全变量, 定

义如
$$v(\mu, X) = \sup \sum_{i=1}^{n} |\mu(E_i)|,$$

这里的上确界 sup 展开于 $B(X)$ 中一切有限个互不相交集 $E_1, \cdots, E_n$ 之上. 在范数 $||\mu|| = v(\mu, X)$ 之下, $B(X)$ 上一切正则可数加有界集合函数 μ 所成的线性空间自然形成一 Banach 空间, 记作 $R(X)$. 记 X 上一切有界连续函数 f 所成 Banach 空间为 $C(X)$, 其范数为 $||f|| = \sup_{x \in X}|f(x)|$. 于是由 Riesz 表示定理, $R(X)$ 与 $C(X)$ 的共轭空间 $C^*(X)$ 在对应 $\mu \leftrightarrow x^*$ 下同构, 这里

$$x^*(f) = \int_X f(x)\mu(dx).$$

最后一式亦将简记为 $\mu(f)$ 或 $f(\mu)$. 命 $R^\omega(X)$ 为与 $R(X)$ 同一集合但在 $C(X)$ 拓扑下的拓扑空间, 其基由以下集合所构成:

$$N(\mu; A, \varepsilon) = \{v/|f(\mu) - f(v)| < \varepsilon, f \in A\},$$

这里 $\mu \in R(X), A \subset C(X)$ 有限, 而 $\varepsilon > 0$ 都任意. 我们知道 $R^\omega(X)$ 是一个局部凸的 Hausdorff 线性拓扑空间 (参阅例如 [2] V. 3).

下述断言, 虽很简单, 但在本文中经常用到, 故仍明确表达如下.

$R^\omega(X)$ 中任一子集 C, 设其对 Banach 空间 $R(X)$ 的线性构造而言是凸的, 则必可缩成一点, 且对 $R^\omega(X)$ 的拓扑构造而言, 这个收缩是连续的.

证之如下: 设 C 为 $R(X)$ 中的子凸集而 μ_0 为 C 的一个定点. 对任意 $\mu \in C$ 与 $0 \leqslant t \leqslant 1$ 命 $\mu_t = t\mu + (1-t)\mu_0 \in C$, 这里 $\mu_1 = \mu$. 定义 $C \times [0,1]$ 到 C 的一个映象 h 为 $h(\mu, t) = \mu_t, \mu \in C, t \in [0,1]$. 置 $h_t : C \to C$ 为 $h_t(\mu) = h(\mu, t)$, 于是 h_t 将 C 收缩成点 μ_0. 为证收缩在 $R^\omega(X)$ 的拓扑下连续, 即 h 对 μ, t 连续, 试考察一固定的 (μ, t) 与 μ_t 在 $R^\omega(X)$ 中的一个邻域 $N = N(\mu_t; A, \varepsilon) = \{v/|f(v) - f(\mu_t)| < \varepsilon, f \in A\}$. 命 $M > 0$ 为 f 取有限集 A 中一切函数时, 较 $|f(\mu) - f(\mu_0)|$ 的最大值为大的一个数. 试考察 (μ, t) 在 $R^\omega(X) \times [0,1]$ 中的一个邻域 U 如下:

$$U = N' \times J, \quad N' = N\left(\mu; A, \frac{\varepsilon}{2}\right) = \left\{v/|f(v) - f(\mu)| < \frac{\varepsilon}{2}, f \in A\right\},$$

$$J = \left\{t'/|t' - t| < \frac{\varepsilon}{2M}, t' \in [0,1]\right\}.$$

对任意 $(v, t') \in U$, 即有

$$f(v_{t'}) - f(\mu_t) = t'f(v) + (1-t')f(\mu_0) - tf(\mu) - (1-t)f(\mu_0)$$
$$= (t' - t)[f(\mu) - f(\mu_0)] + t'[f(v) - f(\mu)].$$

因之

$$|f(v_{t'}) - f(\mu_t)| \leqslant |t' - t| \cdot |f(\mu) - f(\mu_0)| + t'|f(v) - f(\mu)|$$
$$\leqslant \frac{\varepsilon}{2M} \cdot M + \frac{\varepsilon}{2} = \varepsilon,$$

而 $v_{t'} = h(v, t') \in N$. 这证明了 h 在 (μ, t) 连续, 因而, 收缩 h 在 $R^\omega(X)$ 的拓扑下连续.

试考察 X 中的一个子集 F, 并以 $m(F)$ 表 $B(X)$ 上支柱 $[\mu] \subset F$ 的一切正则概率测度 μ 所成的集合. 我们将赋予 $m(F)$ 以拓扑, 使之如拓扑空间 $R^\omega(X)$ 的子空间.

下述引理由定义与上述断言甚显然.

引理 1 (i) 对 X 的任意子集 F, 集合 $m(F)$ 是一凸集 (对 $R(X)$ 的线性构造而言), 因而可连续地缩成一点 (对 $R^\omega(X)$ 的拓扑而言).

(ii) 对 X 的任两子集 F_1, F_2 有

$$m(F_1 \cap F_2) = m(F_1) \cap m(F_2).$$

引理 2 若 F 是 X 的闭集, 则 $m(F)$ 是 $m(X)$ 的闭集.

证. 对任意 $x \bar{\in} F$, 试取含 x 的开集 U_x, V_x, 使

$$x \in U_x \subset \overline{U}_x \subset V_x \subset X - F.$$

依 Urysohn 引理有 X 上的连续函数, 或 $f \in C(X)$, 使 f 在 $X - V_x$ 上 $= 0$, 在 $\overline{U}_x$ 上 $= 1$, 而在 X 上 $0 \leqslant f \leqslant 1$.

今考察任意 $\mu \in \overline{m(F)} \cap m(X)$, 这里记号 $\overline{}$ 指对拓扑空间 $R^\omega(X)$ 而言的闭包. 对任意 $\varepsilon > 0$, 命 $N(\mu; f, \varepsilon)$ 为 $R^\omega(X)$ 中 μ 的下述邻域:

$$N(\mu; f, \varepsilon) = \{v/|f(\mu) - f(v)| < \varepsilon\}.$$

于是有一 $v \in m(F) \cap N(\mu; f, \varepsilon)$, 使 $|f(\mu) - f(v)| < \varepsilon$. 但

$$f(v) = v(f) = \int_X f(x) v(dx) \leqslant v(V_x) = 0.$$

因之 $f(v) = 0$ 而

$$\mu(U_x) = \int_{U_x} f(x)\mu(dx) \leqslant \int_X f(x)\mu(dx) = \mu(f) = f(\mu) < \varepsilon.$$

因 $\varepsilon > 0$ 是任意的, 故有 $\mu(U_x) = 0$, 而 $x \bar{\in} [\mu]$. 因 $x \bar{\in} F$ 是任意的, 故有 $[\mu] \subset F$ 或 $\mu \in m(F)$. 这证明了 $m(F)$ 是 $m(X)$ 的闭集而得本引理.

引理 3 $m(X)$ 是 $R^\omega(X)$ 的闭集.

证. 设 $\mu \in \overline{m(X)}$, 这里记号 $\overline{}$ 指拓扑空间 $R^\omega(X)$ 中的闭包. 于是引理相当于证 μ 是 $B(X)$ 上的一个正则概率测度, 或证明以下二点已足: (i) $\mu(E) \geqslant 0$, 此处 $E \in B(X)$ 任意; (ii) $\mu(X) = 1$.

为证 (i), 试先设其反面 $\mu(E) < 0$, 这里 E 是 $B(X)$ 中的某一闭集. 因 μ 是正则的, 故有一开集 $U \supset E$, 使

$$v(\mu, U - E) < \frac{1}{2}|\mu(E)|$$

(参阅例如 [2] III 5.11 与 III 1.5). 依 Urysohn 引理有一 $f \in C(X)$, 使在 E 上 $f = 1$, 在 $X - U$ 上 $f = 0$, 而在 X 上 $0 \leqslant f \leqslant 1$. 于是有

$$f(\mu) = \int_X f\mu(dx) = \int_E \mu(dx) + \int_{U-E} f\mu(dx)$$
$$\leqslant \mu(E) + \int_{U-E} |f|v(\mu, dx)$$
$$\leqslant \mu(E) + v(\mu, U - E)$$
$$< -\frac{1}{2}|\mu(E)| < 0.$$

试取 $R^\omega(X)$ 中 μ 的一个邻域 N 如下:

$$N = N(\mu; f, \varepsilon) = \{v/|f(\mu) - f(v)| < \varepsilon\},$$

这里 $0 < \varepsilon < |f(\mu)|$. 因 $\mu \in \overline{m(X)}$, 故有 $v \in m(X) \cap N$. 于是

$$v(E) = \int_E fv(dx) \leqslant \int_X fv(dx) = f(v)$$
$$< f(\mu) + \varepsilon < 0,$$

但这与 $v \in m(X)$ 相违, 因之与 $v(E) \geqslant 0$ 相违.

由此知对 $B(X)$ 中任意闭集 E 有 $\mu(E) \geqslant 0$. 设 $E \in B(X)$ 非闭集而 $\mu(E) < 0$. 因 μ 正则, 故依 [2] III 5.11 有一开集 $U \supset E$ 与一闭集 $W \subset E$, 使 $|\mu(C)| < \frac{1}{2}|\mu(E)|$, 这里 $C \in B(X)$ 任意, 只须 $C \subset U - W$. 特别有 $|\mu(E - W)| < \frac{1}{2}|\mu(E)|$, 因而 $\mu(W) = \mu(E) - \mu(E - W) < 0$. 但因 W 是闭的, 前已知其不可能. 于是 (i) 得证.

为证明 (ii) 试设其反 $\mu(X) \neq 1$. 试取 $\varepsilon > 0$ 且 $< |1 - \mu(X)|$. 试考察 X 上函数 $f \equiv 1$ 与 μ 的下述邻域 N:

$$N = N(\mu; f, \varepsilon) = \{v/|f(\mu) - f(v)| < \varepsilon\}.$$

于是有一 $v \in m(X) \cap N$, 使

$$v(X) = f(v) < f(\mu) + \varepsilon = \mu(X) + \varepsilon < 1, \quad \text{在 } \mu(X) < 1 \text{ 时},$$

而

$$v(X) = f(v) > f(\mu) - \varepsilon = \mu(X) - \varepsilon > 1, \text{ 在 } \mu(X) > 1 \text{ 时},$$

不论何时都与 $v \in m(X)$, $v(X) = 1$ 相违. 这证明了 (ii).

至此引理证毕.

命 W 为拓扑空间 $R(X)$ 中的闭单位球体:

$$W = \{\mu/\|\mu\| = v(\mu, X) \leqslant 1, \mu \in R(X)\}.$$

根据 Aloaglu 的一个定理 (参阅例如 [2]V4. 2), 视 W 为拓扑空间 $R^\omega(X)$ 的一个子集时, W 是一复紧集. 因 $m(X) \subset W$, 而 $m(X)$ 是 $R^\omega(X)$ 的闭集 (引理 3), 故 $m(X)$ 也是 $R^\omega(X)$ 的复紧子集. 因对任意 X 的闭集 F, $m(F)$ 是 $m(X)$ 的闭集, 故又有下面的

定理 对 X 的任意闭集 F, 集 $m(F)$ 是拓扑空间 $R^\omega(X)$ 中的复紧闭集.

§4. 支柱从属于覆盖的概率测度集

设 X 为 Hausdorff 复紧空间, 而 $B(X)$ 为 X 中一切 Borel 集所成 σ-域如前. 对 X 的任意子集 F 在 §3 中已定义 $m(F)$ 为 $B(X)$ 上使支柱 $[\mu] \subset F$ 的一切正则概率测度 μ 所成的集合, 并视之为拓扑空间 $R^\omega(X)$ 的子空间. 今考 X 的一个有限闭覆盖 $\mathcal{F} = \{F_1, \cdots, F_r\}$, 由闭集 $F_i, 1 \leqslant i \leqslant r$ 所构成. 定义 $m(\mathcal{F})$ 为 $B(X)$ 上支柱至少全含于诸闭集 $F_i, 1 \leqslant i \leqslant r$ 中之一的一切正则概率测度的集合, 即

$$m(\mathcal{F}) = \sum_{i=1}^{r} m(F_i),$$

并仍视之为 $R^\omega(X)$ 的一个子空间.

定理 对于 X 的有限闭覆盖 $\mathcal{F} = \{F_1, \cdots, F_r\}$, 空间 $R^\omega(X)$ 的子空间 $m(\mathcal{F})$ 具有以下诸性质:

(i) $m(\mathcal{F})$ 是 $R^\omega(X)$ 的闭集.

(ii) $m(\mathcal{F})$ 是 $R^\omega(X)$ 的复紧集.

(iii) $\{m(F_1), \cdots, m(F_r)\}$ 是空间 $m(\mathcal{F})$ 在 Leray 意义下的一个闭凸型覆盖[4].

(iv) 若覆盖 $\mathcal{F}$ 的神经复合形 $K(\mathcal{F})$ 是连通的, 则 $m(\mathcal{F})$ 是 Leray 意义下的凸型空间 [4].

证. 性质 (i), (ii) 直接得自 §3 中的引理 2, 3 与定理. 为证 (iii) 与 (iv), 试先重述 Leray 的若干定义如下.

按 Leray, Hausdorff 复紧空间的一个覆盖称为是凸型的, 如果它满足以下性质:

(a) 覆盖中的每一集合是闭的且是"简单"的, 后者意指具有与一个点相同的 Cêch-Alexander 上同调结构.

(b) 覆盖中任意有限多个集合的交或则是空的, 或则是简单的.

又依 Leray, 一个空间是凸型空间, 如果它是 Hausdorff 复紧的连通空间, 且具有一个凸型覆盖, 除满足以上 (a) 与 (b) 外并满足下述性质 (c):

(c) 对空间的任一点 x 与 x 的任一邻域 V, 在覆盖中必有一集合 U 包含于 V 而含 x 为其内点.

由此定义与 §3 的引理 1 即得推断 (iii). 为证明 (iv), 先注意 $m(\mathcal{F})$ 是 Hausdorff 复紧空间, 且因 $K(\mathcal{F})$ 连通而 $m(\mathcal{F})$ 也是连通的. 今考察任意 $\mu \in m(\mathcal{F})$. 设 F_{i_1}, $\cdots$, F_{i_k} 为覆盖中包含 μ 的支柱 $[\mu]$ 的集合全体. 因每一 $m(F_i)$ 都是 $R^\omega(X)$ 的闭集, 故在 $R^\omega(X)$ 中有 μ 的邻域与所有 $i \neq i_1, \cdots, i_k$, 的 $m(F_i)$ 不相遇. 因 $R^\omega(X)$ 已知是局部凸的, 故在这些邻域中必然有凸的存在. 命 $u(\mu)$ 为 $R^\omega(X)$ 中与一切 $m(F_i), i \neq i_1, \cdots, i_k$, 不相遇的闭凸邻域的全体. 又命 $U(\mu)$ 为 $m(\mathcal{F})$ 中由 $u(\mu)$ 中集合与 $m(\mathcal{F})$ 的交所构成一切子集的全体. 于是对于一切 $\mu \in m(\mathcal{F})$, 所有 $U(\mu)$ 中集合的全体 U 构成空间 $m(\mathcal{F})$ 的一个闭覆盖具有使 $m(\mathcal{F})$ 成为凸型空间的性质 (a), (b) 与 (c). (c) 的成立是由于 $U(\mu)$ 构成 μ 在 $m(\mathcal{F})$ 中的一个邻域系统, (a) 是因为 U 中每一集合 V 都是闭凸因而也是简单的, 而 (b) 是因为任意有限多个闭凸集的交, 必然也是闭且凸的, 因而如果不是空的, 就是简单的.

§5. 前述诸概念的推广

设 X 与 $B(X)$ 如前. 命 c 为任意定数. 对 X 的任意集合 $F \in B(X)$ 将以 $m_c(F)$ 表 $B(X)$ 上使 $\mu(F) \geqslant c$ 的所有正则概率测度 μ 所成的集合. 在 $c > 1$ 时, $m_c(F)$ 显为空集. 在 $c = 1$ 时 $m_1(F)$ 即为 §3 中的集合 $m(F)$. 在 $c \leqslant 0$ 时, $m_c(F)$ 与 $m(X)$ 相同. 在一般情形, $m_c(F)$ 可解说为 $B(X)$ 上使 $\mu([\mu] \cap F) \geqslant c$ 的一切正则概率测度所成的集合.

引理 1 集合 $m_c(F)$ 具有以下诸性质:

(i) 在 $c \leqslant 1$ 时, $m_c(X)$ 与 $m(X)$ 相同, 因而是空间 $R^\omega(X)$ 中闭复紧集.

(ii) 在 $d < c \leqslant 1$ 时, $m(F) \subset m_c(F) \subset m_d(F) \subset m(X)$.

(iii) 在 $c \leqslant 1$ 时, $m_c(F)$ 对 $R(X)$ 的线性构造而言是凸的, 且在 $R^\omega(X)$ 的拓扑下可连续地缩成一点.

(iv) 对任意 c_1, c_2, F_1, F_2 有

$$m_{c_1}(F_1) \cap m_{c_2}(F_2) \subset m_{c_1+c_2-1}(F_1 \cap F_2).$$

证. 由定义直接得出 (参阅 §3).

引理 2 (i) $m_c(F)$ 在空间 $R^\omega(X)$ 中的闭包 $\overline{m_c(F)}$ 为包含于 $m(X)$ 中 $R^\omega(X)$ 的一个复紧集.

(ii) 若 F 是 X 的闭集, 则 $m_c(F)$ 是空间 $R^\omega(X)$ 中的闭复紧集.

证. 因 $m_c(F) \subset m(X)$ 且由 §3 中引理 3 与定理 $m(X)$ 是 $R^\omega(X)$ 中的闭复紧集, 故 $\overline{m_c(F)} \subset m(X)$ 且是 $R^\omega(X)$ 中的复紧集. 这证明了 (i). 为证明 (ii) 设 $0 < c < 1$, F 是 X 的闭集而 $v \in \overline{m_c(F)}$. 于是由 (i), v 是一正则概率测度. 若 $v \bar\in m_c(F)$, 则有 $v(F) < c < 1$, 因而 $[v] \not\subset F$, 且 $v([v] - F) > 1 - c > 0$. 因 v 正则, 故有闭集 $C \subset [v] - F$ 使

$$v([v] - F - C) < v([v] - F) - 1 + c.$$

于是有

$$v(C) > 1 - c > 0.$$

因 C 与 F 都是闭集且不相遇, 故有 X 上的连续函数 f 使在 C 上 $f = 1$, 在 F 上 $f = 0$, 而在 X 上 $0 \leqslant f \leqslant 1$. 置 $\varepsilon = v(c) - 1 + c > 0$, 并考虑 v 在 $R^\omega(X)$ 中的下述邻域:

$$N = N(v; f, \varepsilon) = \{\mu / |f(\mu) - f(v)| < \varepsilon\}.$$

因 $v \in \overline{m_c(F)}$, 故有 $\mu \in m_c(F) \cap N$. 对此 μ 有

$$\begin{aligned}
\mu(X - F) &\geqslant \int_X f(x) \mu(dx) = f(\mu) \\
&> f(v) - \varepsilon = \int_X f(x) v(dx) - \varepsilon \\
&\geqslant \int_C f(x) v(dx) - \varepsilon = v(C) - \varepsilon \\
&= 1 - c.
\end{aligned}$$

于是 $\mu(F) < c$, 而 $\mu \bar\in m_c(F)$ 与 μ 的选取相违. 这证明了 (ii) 在 $0 < c < 1$ 时成立. 在 $c \geqslant 1$ 或 $c \leqslant 0$ 的情形则甚显然.

今考察 X 的一个闭覆盖 $\mathcal{F} = \{F_1, \cdots, F_r\}$ 与一组数 $c = \{c_1, \cdots, c_r\}$. 我们将置

$$m_c(\mathcal{F}) = \sum_{i=1}^{r} m_{c_i}(F_i),$$

并视之为空间 $R^\omega(X)$ 的子空间.

定理 设对每一 $i = 1, \cdots, r$, 数 $c_i \geqslant 0$ 且 $\leqslant 1$. 则 $R^\omega(X)$ 的子空间 $m_c(\mathcal{F})$ 具有以下诸性质:

(i) $m_c(\mathcal{F})$ 是 $R^\omega(X)$ 的闭集.

(ii) $m_c(\mathcal{F})$ 是 $R^\omega(X)$ 的复紧集.

(iii) $C_c(\mathcal{F}) = \{m_{c_1}(F_1), \cdots, m_{c_r}(F_r)\}$ 是 $m_c(\mathcal{F})$ 在 Leray 意义下的一个闭凸型覆盖.

(iv) 若 $m_c(\mathcal{F})$ 由集合 $m_{c_i}(F_i), 1 \leqslant i \leqslant r$ 所构成的覆盖 $C_c(\mathcal{F})$ 的神经复合形 $K_c(\mathcal{F})$ 是连通的, 则 $m_c(\mathcal{F})$ 是 Leray 意义下的凸型空间.

(v) 若对从 1 至 r 取出的任一组指数 $i_1, \cdots, i_s$ 有

$$c_{i_1} + \cdots + c_{i_s} > s - 1$$

$\left(\text{特别在 } c_i > 1 - \dfrac{1}{r}, 1 \leqslant i \leqslant r \text{ 时}\right)$, 则 $m_c(\mathcal{F})$ 的覆盖 $C_c(\mathcal{F})$ 的神经复合形 $K_c(\mathcal{F})$ 与 X 的覆盖 $\mathcal{F}$ 的神经复合形 $N(\mathcal{F})$ 同构.

证. 关于 (i)—(iv) 证明与 §4 的引理相仿. 为证明 (v), 试先注意在对应 $F_i \leftrightarrow m_{c_i}(F_i), 1 \leqslant i \leqslant r$ 下 $N(\mathcal{F})$ 可视为 $K_c(\mathcal{F})$ 的一个子复合形. 试考察任意一组指数 $i_1, \cdots, i_s$, 对此有

$$F_{i_1} \cap \cdots \cap F_{i_s} = \varnothing,$$
$$F_{i_1} \cap \cdots \cap \hat{F}_{i_j} \cap \cdots \cap F_{i_s} \neq \varnothing, \quad 1 \leqslant j \leqslant s.$$

这里记号 $\hat{F}_{i_j}$ 指集合 F_{i_j} 不估计在相应的交以内. 假设

$$m_{c_{i_1}}(F_{i_1}) \cap \cdots \cap m_{c_{i_s}}(F_{i_s}) \neq \varnothing,$$

而 μ 在此交集中, 则由引理 1(iv) 有

$$\mu \in m_{c_{i_1}}(F_{i_1}) \cap \cdots \cap \hat{m}_{c_{i_j}}(F_{i_j}) \cap \cdots \cap \hat{m}_{c_{i_s}}(F_{i_s})$$
$$\subset m_d(F_{i_1} \cap \cdots \cap \hat{F}_{i_j} \cap \cdots \cap F_{i_s}),$$

此处

$$d = c_{i_1} + \cdots + c_{i_s} - c_{i_j} - s + 2.$$

由此得

$$\mu([\mu] \cap F_{i_1} \cap \cdots \cap \hat{F}_{i_j} \cap \cdots \cap F_{i_s}) \geqslant c_{i_1} + \cdots + c_{i_s} - c_{i_j} - s + 2.$$

因诸集合 $F_{i_1} \cap \cdots \cap \hat{F}_{i_j} \cap \cdots \cap F_{i_s}$ 互不相遇，故有

$$1 = \mu([\mu]) \geqslant \sum_{j=1}^{s} \mu([\mu] \cap F_{i_1} \cap \cdots \cap \hat{F}_{i_j} \cap \cdots \cap F_{i_s})$$

$$\geqslant \sum_{j=1}^{s}(c_{i_1} + \cdots + c_{i_s} - c_{i_j} - s + 2)$$

$$= (s-1)(c_{i_1} + \cdots + c_{i_s}) - s(s-2).$$

由此得 $c_{i_1} + \cdots + c_{i_s} \leqslant s - 1$ 与假设相违。这证明了 $K_c(\mathcal{F})$ 与 $N(\mathcal{F})$ 同构。

§6. 多值映象的一致闭性

设 T 是 Hausdorff 复紧空间 X 到 Hausdorff 复紧空间 Y 的一个多值映象。积空间 $X \times Y$ 中由一切使 $y \in T(x)$ 的 (x,y) 所成的子集称为 T 的图形，并将以 $G(T)$ 表之。如果 $G(T)$ 是 $X \times Y$ 的闭集，则 T 称为闭的。闭映象 T 将称为一致闭的，如果对任意 $(x,y) \in G(T)$ 与 y 在 Y 中的任意邻域 V 有一 x 在 X 中的邻域 U 使对任意 $x' \in U$，集合 $T(x') \cap V$ 非空。

引理 设 Hausdorff 复紧空间 X 到 Hausdorff 复紧空间 Y 的多值映象 T 是闭的也是一致闭的，则对任意 $X \times Y$ 上的连续函数 f，由

$$T_f(x) = \{y/y \in T(x), f(x,y) = \sup_{\overline{y} \in T(x)} f(x,\overline{y})\} \subset T(x)$$

所定义的 X 到 Y 中的映象 T_f 是一个闭映象。

证. 置 $\sup\limits_{\overline{y} \in T(x)} f(x,\overline{y}) = m_x, x \in X$. 设 $(x,y) \in \overline{G(T_f)}$ 而 $y_0 \in T_f(x)$ 使 $f(x,y_0) = m_x$. 因 $G(T)$ 是闭的，故 $(x,y) \in G(T)$. 若 $(x,y) \overline{\in} G(T_f)$，则 $f(x,y) < m_x$. 置 $\varepsilon = m_x - f(x,y) > 0$，而设 U, V, V_0 各为 x, y, y_0 在 X 与 Y 中的邻域使对任意 $x' \in U, y' \in V, y_0' \in V_0$ 有

$$|f(x',y') - f(x,y)| < \frac{\varepsilon}{2},$$

与

$$|f(x',y_0') - f(x,y_0)| < \frac{\varepsilon}{2}.$$

因 T 是一致闭的，故有 x 在 X 中的邻域 $W \subset U$ 使对任意 $x' \in W$ 有 $T(x') \cap V \neq \varnothing$ 与 $T(x') \cap V_0 \neq \varnothing$. 因 $(x,y) \in \overline{G(T_f)}$，故有 $(\bar{x}', \bar{y}') \in G(T_f)$ 使 $\bar{x}' \in W, \bar{y}' \in V$. 对此 $\bar{x}'$ 而言又有一 $\bar{y}_0' \in V_0 \cap T(\bar{x}')$. 于是有

$$f(\bar{x}', \bar{y}') = m_{\bar{x}'} \geqslant f(\bar{x}', \bar{y}_0').$$

另一面又有

$$\begin{aligned}
f(\bar{x}', \bar{y}_0') &= f(\bar{x}', \bar{y}') + [f(x,y) - f(\bar{x}', \bar{y}')] \\
&\quad + [f(x, y_0) - f(x,y)] + [f(\bar{x}', \bar{y}_0') - f(x, y_0)] \\
&> f(\bar{x}', \bar{y}') - \frac{\varepsilon}{2} + \varepsilon - \frac{\varepsilon}{2} \\
&= f(\bar{x}', \bar{y}').
\end{aligned}$$

与前式相违. 因之 $(x,y) \in G(T_f)$ 或 $G(T_f)$ 是 $X \times Y$ 的闭集或即 T_f 是闭的.

§7. 关于多值映象的几个拓扑定理

对于 Hausdorff 复紧空间 X 将以 $H(X)$ 表在有理数域上的 Cêch-Alexander 上同调环. 空间将称为简单的 (或更准确地说对于有理数域的系数域来说是简单的), 如果它与一个点有相同的上同调环.

在以后将用到下面两个一般的引理.

引理 1 (Leray)[4]　若 Hausdorff 复紧空间 X 有一在 Leray 意义下有限的闭凸型覆盖 $\mathcal{F}$, 则空间 X 与 $\mathcal{F}$ 的神经复合形 N 有同样的上同调环: $H(X) \approx H(N)$. 特别有 $\chi(X) = \chi(N)$.

引理 2 (Vietoris-Begle)[1]　若 f 是 Hausdorff 复紧空间 X 到 Hausdorff 复紧空间 Y 的一个连续映象 f, 使对任意 $y \in Y, f^{-1}(y)$ 是简单的, 则映象 f 引出同构 $f^* : H(Y) \approx H(X)$.

今设 φ, ψ 是 Hausdorff 复紧空间 X 到凸型空间 Y 的两个连续映象, 这里对任意 $y \in Y, \varphi^{-1}(y)$ 都是简单的. 因 Y 是凸型的, 故上同调环 $H(Y)$ 有一有限基设为 $Z_i^p, 0 \leqslant p \leqslant N, 1 \leqslant i \leqslant \alpha_p, p$ 为相应维数. 由上面的 Vietoris-Begle 定理, $H(X)$ 也有一有限基由 $\varphi^*(Z_i^p)$ 构成, 这里 $\varphi^* : H(Y) \to H(X)$ 是 φ 引出的同构. 由此得

$$\psi^*(Z_i^p) = \sum_j b_{ij}^p \varphi^*(Z_j^p).$$

以 $S_p B^p$ 表矩阵 $B^p = (b_{ij}^p)$ 的迹, 则数 $\sum_p (-1)^p S_p B^p$ 与基 $\{Z_i^p\}$ 的选择无关而将记之为 $\Lambda(\varphi, \psi)$.

定理 A　设 φ, ψ 为 Hausdorff 复紧空间 X 到凸型空间 Y 的两个连续映象并设对任意 $y \in Y, \varphi^{-1}(y)$ 都是简单的. 若 $\Lambda(\varphi, \psi) \neq 0$, 则 φ, ψ 有一重合点, 即有点 $x \in X$ 使 $\varphi(x) = \psi(x)$.

上述定理与 Leray 关于映象的定点定理相仿, 其证明将从略. 下一定理则由定义直接得出.

定理 B　设 φ 是 Hausdorff 复紧空间 X 到 Hausdorff 复紧空间 Y 的一个连续映象，并设对任意 $y \in Y, \varphi^{-1}(y)$ 都是简单的. 于是 $\Lambda(\varphi, \varphi) = \chi(Y)$，这里 $\chi(Y)$ 指 Y 的 Euler-Poincaré 示性数.

今设 T 是凸型空间 Y 到自身的一个多值映象，具有以下二性质：

(i) T 是闭的；

(ii) 对任意 $y \in Y, T(y)$ 都是简单的.

今记 T 的图形 $G(T)$ 为 X，并依 $Y \times Y$ 到 Y 的两个投影定义 X 到 Y 的两个映象如下：

$$\begin{aligned}\varphi(y, y') &= y, \\ \psi(y, y') &= y'\end{aligned} \quad (y' \in T(y) \text{或} (y, y') \in X).$$

由 (i)，T 的图形 $X = G(T)$ 是 $Y \times Y$ 的闭集，因之 X 是 Hausdorff 复紧空间. 因对每一 $y \in Y, \varphi^{-1}(y)$ 与 $T(y)$ 在 ψ 下同拓而依 (ii) 是简单的，故数 $\Lambda(\varphi, \psi)$ 有定义. 我们定义

$$\Lambda(T) = \Lambda(\varphi, \psi).$$

定理 C　设 T 是凸型空间 Y 到自身的一个多值闭映象，并设对每一 $y \in Y$，$T(y)$ 都是简单的. 若数 $\Lambda(T) \neq 0$，则 T 有一定点，即有一点 $y \in Y$ 使 $y \in T(y)$.

证.　定义 $X = G(T)$ 与映象 $\varphi, \psi: X \to Y$ 如前. 因 $\Lambda(\varphi, \psi) = \Lambda(T) \neq 0$，故依定理 A，$\varphi, \psi$ 有重合点 $x = (y, y') \in X$ 使 $\varphi(x) = \psi(x)$，亦即 $y = y' \in T(y)$. 证毕.

定理 D　设 T 是凸型空间 Y 到自身的恒同映象，则 $\Lambda(T) = \chi(Y)$.

证.　这由定理 B 直接得出.

定理 E　设 T_0, T_1 是凸型空间 Y 到自身的两个多值闭映象，并设

(i) 有一 $\widetilde{Y} = Y \times [0, 1]$ 到 $\widetilde{Y}$ 的多值闭映象 $\widetilde{T}$，使 $\widetilde{T}(y, k) = T_k(y)$，这里 $k = 0, 1, y \in Y$，而 $\widetilde{T}(Y \times (t)) \subset Y \times (t), t \in [0, 1]$.

(ii) 定义 $T_t: Y \to Y$ 为 $(T_t(y), t) = \widetilde{T}(y, t), t \in [0, 1]$，则对每一 $y \in Y$ 与 $t \in [0, 1], T_t(y)$ 都是简单的. 于是 $\Lambda(T_0) = \Lambda(T_1)$.

证.　命 $\widetilde{X} = G(\widetilde{T}), X_0 = G(T_0), X_1 = G(T_1)$ 各为 $\widetilde{T}, T_0$ 与 T_1 的图形. 定义投影 $\widetilde{\varphi}, \widetilde{\psi}: \widetilde{X} \to \widetilde{Y}, \varphi_0, \psi_0: X_0 \to Y_0 = Y \times (0)$ 与 $\varphi_1, \psi_1: X_1 \to Y_1 = Y \times (1)$，如

$$\widetilde{\varphi}(\widetilde{y}, \widetilde{y}') = \widetilde{y}, \quad \widetilde{\psi}(\widetilde{y}, \widetilde{y}') = \widetilde{y}', \quad \varphi_k(y_k, y_k') = y_k, \quad \psi_k(y_k, y_k') = y_k',$$

这里 $(\widetilde{y}, \widetilde{y}') \in \widetilde{X}, (y_k, y_k') \in X_k, k = 0, 1$. 记 Y 到 $\widetilde{Y}$ 中 $Y_k = Y \times (k)$ 的自然映入为 λ_k，即

$$\lambda_k(y) = (y, k), \quad y \in Y, \quad k = 0, 1.$$

同样记 X_k 到 $\widetilde{X}$ 的自然映入为 θ_k, $k = 0, 1$. 任取 $H(\widetilde{Y})$ 的一个基 $\{\widetilde{Z}_i^p\}$, 则 $\{\lambda_k^* \widetilde{Z}_i^p\}$, $k = 0, 1$, 都是 $H(Y)$ 的基. 今有

$$\Lambda(\widetilde{T}) = \Lambda(\widetilde{\varphi}, \widetilde{\psi}) = \sum_p (-1)^p S_p(b_{ij}^p),$$

这里

$$\widetilde{\psi}^*(\widetilde{Z}_i^p) = \sum_j b_{ij}^p \widetilde{\varphi}^*(\widetilde{Z}_j^p).$$

因 $\widetilde{\varphi}\theta_k = \lambda_k \varphi_k, \widetilde{\psi}\theta_k = \lambda_k \psi_k$, 故应用 θ_k^* 于以上方程的两边得

$$\psi_k^*(\lambda_k^* \widetilde{Z}_i^p) = \sum_j b_{ij}^p \varphi_k^*(\lambda_k^* \widetilde{Z}_j^p), \quad k = 0, 1.$$

由此得

$$\Lambda(T_k) = \Lambda(\varphi_k, \psi_k) = \sum_p (-1)^p S_p(b_{ij}^p),$$

故 $\Lambda(T_0) = \Lambda(T_1) = \Lambda(\widetilde{T})$, 而定理得证.

注. 为简单起见, 我们称满足上定理中条件的两个映象为简单同伦的.

§8. 对策的定义与主要定理

试考察一 n 人对策, 第 i 人的策略空间为 S_i, 而赢得函数为 $H_i(x_1, \cdots, x_n)$, $x_i \in S_i$, $i = 1, \cdots, n$. 我们将设 S_i 都是 Hausdorff 复紧空间, 而 H_i 都是 $S = S_1 \times \cdots \times S_n$ 上的连续函数. 对每一 S_i 设 $\mathcal{F}_i = \{F_1^{(i)}, \cdots, F_{m_i}^{(i)}\}$ 是它的一个有限闭覆盖, B_i 是 S_i 上一切 Borel 集所成的 σ-域, 而 $\{c_i\} = \{c_{i1}, \cdots, c_{im_i}\}$ 是一组 $\geqslant 0$ 且 $\leqslant 1$ 的数. 如 §5 所定义, 命 $S_i^* = m_{c_i}(\mathcal{F}_i)$ 为 B_i 上使至少对一指数 j, $1 \leqslant j \leqslant m_i$, 有 $\mu_i(F_j^{(i)}) \geqslant c_{ij}$ 的一切正则概率测度 μ_i 的集合, 具有由拓扑空间 $R^\omega(S_i) = R_i^\omega$ 所引出的拓扑.

今对每一 $i = 1, \cdots, n$, 试考察一 S_i^* 到自身的多值映象 τ_i, 具有以下诸性质:

(i) $\mu_i \in \tau_i(\mu_i)$, $\mu_i \in S_i^*$.

(ii) τ_i 是闭的也是一致闭的.

(iii) 对每一 $\mu_i \in S_i^*$, 集 $\tau_i(\mu_i)$ 对 Banach 空间 $R_i = R(S_i)$ 而言的线性构造来说都是凸集.

定义 系统 $\Gamma = \langle I, \{S_i\}, \{H_i\}, \{\mathcal{F}_i\}, \{c_i\}, \{\tau_i\}\rangle$, 其中 $I = \{1, \cdots, n\}$, 为对策者集, 将称为一个活动受限制的对策. 覆盖 $\mathcal{F}_i$ 中的闭集 $F_j^{(i)}$ 将称第 i 人的活动区域, τ_i 为其改变区域, 而 c_{ij} 为其集中度. 考虑对策 $\Gamma^* = \langle I, \{S_i^*\}, \{H_i^*\}, \{\tau_i\}\rangle$, 这里

对策者集同样是 I, 策略空间为 $S_i^* = m_{c_i}(\mathcal{F}_i) = \sum_i m_{c_{ij}}(F_j^{(i)}) \subset R^\omega(S_i)$, 而赢得函数为 $H_i^*(\mu_1, \cdots, \mu_n) = \int_S H_i(x_1, \cdots, x_n)\mu(dx)$, 其中 μ 为积空间 $S = S_1 \times \cdots \times S_n$ 上正则概率测度 $\mu_i \in S_i^*$ 的积测度. 我们将称 Γ^* 为对策 Γ 的自然扩充, 并称 $(\mu_1^*, \cdots, \mu_n^*) \in S_1^* \times \cdots \times S_n^*$ 为 Γ 或 Γ^* 的一个平衡局势, 如果对任意 $\mu_i \in \tau_i(\mu_i^*)$, 有

$$H_i^*(\mu_1^*, \cdots, \mu_i^*, \cdots, \mu_n^*) \geqslant H_i^*(\mu_1^*, \cdots, \mu_i, \cdots, \mu_n^*)$$

$(i = 1, \cdots, n)$. 记空间 $m_{c_i}(\mathcal{F}_i)$ 的覆盖 $\{m_{c_{ij}}(F_{ij})\}$ 的神经复合形为 $K_i = K_{c_i}(\mathcal{F}_i)$, 复合形的 Euler-Poincaré 示性数为 χ_i. 则数 $\chi(\Gamma) = \chi_1 \cdots \chi_n$ 将称为对策 Γ 的示性数.

主要定理 活动受限制的对策 $\Gamma = \langle I, \{S_i\}, \{H_i\}, \{\mathcal{F}_i\}, \{c_i\}, \{\tau_i\}\rangle$ 在所有神经复合形 K_i 都连通且示性数 $\chi(\Gamma) \neq 0$ 时, 必有平衡局势.

证. 对任意 $\mu = (\mu_1, \cdots, \mu_n) \in S^* = S_1^* \times \cdots \times S_n^*$, 命 $\Phi^{(i)}(\mu)$ 为使 $H_i^*(\mu_1, \cdots, \mu_i', \cdots, \mu_n) = \sup\limits_{v_i \in \tau_i(\mu_i)} H_i^*(\mu_1, \cdots, v_i, \cdots, \mu_n)$, 又使 $\mu_i' \in \tau_i(\mu_i) \subset S_i^*$ 的一切 μ_i' 的集合, 又命

$$\Phi(\mu) = \Phi^{(1)}(\mu) \times \cdots \times \Phi^{(n)}(\mu) \subset S^*.$$

因 τ_i 是闭的, 故 $\Phi(\mu)$ 是非空的. 因 τ_i 又是一致闭的, 故由 §6 的引理, Φ 是闭的. 因 $\tau_i(\mu_i)$ 是凸的, 故 $\Phi(\mu)$ 也是凸的 (对 Banach 空间 $R = R(S_1) \times \cdots \times R(S_n)$ 的线性构造而言). 而且, 因为 $\tau_i(\mu_i)$ 是凸的且含有 μ_i, 故 Φ "简单地同伦" 于 S^* 到 S^* 的恒同映象 J. 由此从 §7 的定理 E, D 得

$$\Lambda(\Phi) = \Lambda(J) = \chi(S^*) = \prod_{i=1}^{n} \chi(S_i^*).$$

又由 §7 的引理 1 与 §5 的定理有

$$\chi(S_i^*) = \chi(K_i) = \chi_i.$$

因之

$$\Lambda(\Phi) = \chi_1 \cdots \chi_n = \chi(\Gamma) \neq 0.$$

由 §7 的定理 C, 应有一点 $\mu^* \in S^*$ 使 $\mu^* \in \Phi(\mu^*)$. 这一点 μ^* 即为对策的一个平衡局势而定理得证.

推论 1 若对每一 i 与从 $1, \cdots, m_i$ 取出的指数组 $j_1, \cdots, j_s$ 有 $c_{ij_1} + \cdots + c_{ij_s} > s - 1$, 则活动受限制的对策 $\Gamma = \langle I, \{S_i\}, \{H_i\}, \{(\mathcal{F}_i)\}, \{c_i\}, \{\tau_i\}\rangle$ 恒有平衡局势, 只须诸 Euler-Poincaré 数 $\chi(N_i)$ 无一为 $0, i = 1, \cdots, n$, 这里 N_i 为覆盖 $\mathcal{F}_i$ 的神经复合形, 假定是连通的.

证. 由 §5 的定理中 (v)

$$\chi(N_i) = \chi_i = \chi(K_{c_i}(\mathcal{F}_i)).$$

由此得本推论.

推论 2 若每一 $\mathcal{F}_i$ 仅由一个集合即 S_i 自身所构成, 则活动受限制对策 $\Gamma = \langle I, \{S_i\}, \{H_i\}, \{\mathcal{F}_i\}, \{c_i\}, \{\tau_i\} \rangle$ 或简记为 $\Gamma = \langle I, \{S_i\}, \{H_i\}, \{\tau_i\} \rangle$, 必有平衡局势.

证. 盖在此情形 $K_{c_i}(\mathcal{F}_i)$ 为一个点, 因而 $\chi_i = 1 \neq 0$.

推论 3 (Nash-Glicksberg)[3,5] 设对策 $\Gamma = \langle I, \{S_i\}, \{H_i\} \rangle$ 中的 S_i 都是 Hausdorff 复紧空间而 H_i 都在 $S = S_1 \times \cdots \times S_n$ 上连续, 则 Γ 必有平衡局势.

证. 因对策 Γ 可视为一个活动受限制的对策, 其中每一 $\mathcal{F}_i$ 只由一个集合即 S_i 自身所构成, 每一 $c_i = 1$, 而对每一 $\mu_i \in S_i^*$ 有 $\tau_i(\mu_i) = S_i^*, 1 \leqslant i \leqslant n$.

结论 对于空间 S_i 的一个有限闭覆盖 $\mathcal{F}_i = \{F_j^{(i)}\}, 1 \leqslant j \leqslant m_i$ 而言, 它的神经复合形 N_i 的 Euler-Poincaré 数 $\chi_i = \chi(N_i)$ 等于

$$\chi_i = \sum_{S=0}^{m_i-1} (-1)^s a_s(\mathcal{F}_i),$$

这里 $a_s(\mathcal{F}_i)$ 表从闭集 $F_j^{(i)}$ 中选出无公共交集的一切 $(s+1)$ 组的个数. 因之 χ_i 是一个由覆盖 $\mathcal{F}_i$ 的诸闭集相互间的关系所确定的数. 故推论 1 的断言是: 只须策略的选择充分集中, 且诸活动区域的相互关系满足 $\chi(\Gamma) \neq 0$ 与连通性的数, 即能保证平衡局势的存在. 推论 2 则指出, 如果策略的选择不受任何限制, 则平衡局势必然存在, 而与策略空间的构造及策略的改变区域无关. 如果策略的改变也不受任何限制, 即得 Nash-Glicksberg 定理 (推论 3). 另一面, 简单的例子 (见 §9 的例) 指出, 如果 $\chi(\Gamma) = 0$, 那么即使策略空间很简单以至只含有有限个点, 平衡局势也不必存在. 因之, 我们的定理说明:

决定一个活动受限制对策的平衡局势存在与否的主要因素, 乃是诸活动区域间相互错综复杂的关系, 而非策略空间自身.

§9. 例

定义一只有两个人的活动受限制对策 $\Gamma = \langle I, \{S_i\}, \{H_i\}, \{\mathcal{F}_i\}, \{c_i\}, \{\tau_i\} \rangle$ 如下:

设对策者 II 有 4 个 (纯正) 策略 $a_i, 1 \leqslant i \leqslant 4$, 而对策者 II 有 4 个 (纯正) 策略 $b_j, 1 \leqslant j \leqslant 4$. 赢得函数 H_1 与 H_2 各如下表所示:

H_1	a_1	a_2	a_3	a_4
b_1	γ	β	α	δ
b_2	β	α	δ	γ
b_3	α	δ	γ	β
b_4	δ	γ	β	α

H_2	a_1	a_2	a_3	a_4
b_1	β	γ	δ	α
b_2	γ	δ	α	β
b_3	δ	α	β	γ
b_4	α	β	γ	δ

表中的数 $\alpha, \beta, \gamma, \delta$ 如此选择, 使满足以下诸不等式:

$$\delta < \alpha < \beta < \gamma, \tag{1}$$

$$\alpha < 2\delta, \tag{2}$$

$$\gamma + \delta < 2\alpha, \tag{3}$$

$$\alpha + \gamma < 2\beta. \tag{4}$$

覆盖 $\mathcal{F}_i, i = 1, 2,$ 则各由 4 个闭集 $F_j^{(i)}, 1 \leqslant j \leqslant 4$, 所构成, 这里

$$F_j' = \{a_j, a_{j+1}\},$$

$$F_j'' = \{b_j, b_{j+1}\}$$

(规约 $a_5 = a_1, b_5 = b_1$). 诸数 $\{c_{ij}\}, i = 1, 2$, 将取作都与 > 0 又 < 1 的数 c 相等, 这里的 c 并将假定充分接近于 1. 于是空间 $S_i^*, i = 1, 2$, 可视为各由点 $\sum_{j=1}^{4} x_j a_j$ 与 $\sum_{j=1}^{4} y_j b_j$ 所构成, 这里的 x, y 各满足以下诸不等式:

对 x: $x_j \geqslant 0, 1 \leqslant j \leqslant 4$,

$$\sum_{j=1}^{4} x_j = 1.$$

又 $x_1 + x_2 \geqslant c$ 或 $x_2 + x_3 \geqslant c$ 或 $x_3 + x_4 \geqslant c$ 或 $x_4 + x_1 \geqslant c$.

对 y: $y_j \geqslant 0, 1 \leqslant j \leqslant 4$,

$$\sum_{j=1}^{4} y_j = 1.$$

又 $y_1 + y_2 \geqslant c$, 或 $y_2 + y_3 \geqslant c$, 或 $y_3 + y_4 \geqslant c$, 或 $y_4 + y_1 \geqslant c$.

命 $a_j', a_j'', b_j', b_j'', 1 \leqslant j \leqslant 4$, 各为由以下诸式所定的点:

$$a'_1 = ca_1 + (1-c)a_3,$$
$$a'_2 = (2c-1)a_2 + (1-c)a_1 + (1-c)a_3,$$
$$a'_3 = ca_3 + (1-c)a_1,$$
$$a'_4 = (2c-1)a_4 + (1-c)a_1 + (1-c)a_3,$$
$$a''_1 = (2c-1)a_1 + (1-c)a_2 + (1-c)a_4,$$
$$a''_2 = ca_2 + (1-c)a_4,$$
$$a''_3 = (2c-1)a_3 + (1-c)a_2 + (1-c)a_4,$$
$$a''_4 = ca_4 + (1-c)a_2.$$

同样 b'_j, b''_j 亦由类似的等式所定义, 只是在以上各式中诸 a 都易为相应的 b. 对策者 II 与 II 的所有混合策略所成的空间于是可视为四面体 T_1 与 T_2, 其顶点各为 a_j, $1 \leqslant j \leqslant 4$, 与 $b_j, 1 \leqslant j \leqslant 4$. 于是 S_1^* 是 T_1 在四棱 $a_1a_2, a_2a_3, a_3a_4, a_4a_1$ 周围的某一部分, 这一部分的边界系由四个平行四边形 $a'_1a''_1a'_2a''_2, a'_2a''_2a'_3a''_3, a'_3a''_3a'_4a''_4, a'_4a''_4a'_1a''_1$ 与其他 8 个梯形所构成, 这些梯形两两在四面体 T_1 的四个面上. 我们将以 $C_j, 1 \leqslant j \leqslant 4$, 表 S_1^* 在 a_j 附近的四个角, 对此 C_1 系由以下诸式所定:

$$\begin{cases} x_1 + x_2 \geqslant c, x_1 + x_4 \geqslant c, \\ x_1 \geqslant 0, x_2 \geqslant 0, x_3 \geqslant 0, x_4 \geqslant 0, \\ x_1 + x_2 + x_3 + x_4 = 1. \end{cases}$$

其他诸角 $C_j, j = 2, 3, 4$, 亦类似. 我们又将以 $C_{j,j+1}$, 表由以下诸式所定义的四个柱形:

$$\begin{cases} x_j + x_{j+1} \geqslant c, x_{j-1} + x_j < c, x_{j+1} + x_{j+2} < c, \\ x_1 \geqslant 0, x_2 \geqslant 0, x_3 \geqslant 0, x_4 \geqslant 0, \\ x_1 + x_2 + x_3 + x_4 = 1, \end{cases}$$

这里 $1 \leqslant j \leqslant 4$ 并按规约 $x_{k+4} = x_k, C_{4,5} = C_{4,1}$.

现在再定义对于 $\mu \in S_1^*$ 的改变区域 $\tau_1(\mu)$, 使 τ_1 除满足凸、闭与一致闭以及包含 μ 自身的一些要求外, 并满足以下诸条件:

(i) $\tau_1(\mu) = C_j$, μ 在线段 $a'_j a''_j$ 上时, $1 \leqslant j \leqslant 4$.

(ii) $\tau_1(\mu) \supset C_j$, $\mu \in C_j$ 时, $1 \leqslant j \leqslant 4$.

(iii) $\tau_1(\mu) \subset C_{j,j+1} \cup C_j \cup C_{j+1}$, $\mu \in C_{j,j+1}$ 时, $1 \leqslant j \leqslant 4$.

(iv) $\mu \in \text{int}\tau_1(\mu)$, μ 不在任一线段 $a'_j a''_j$ 上时, $1 \leqslant j \leqslant 4$.

对满足 $u \geqslant 0, u' \geqslant 0, u'' \geqslant 0, u + u' + u'' = 1$ 的数 (u, u', u'') 与满足 $v \geqslant 0$,

$v' \geqslant 0, v'' \geqslant 0, v + v' + v'' = 1$ 的数 (v, v', v'')，命

$$\bar{a}_j = ua_j + u'a'_j + u''a''_j, \quad (1 \leqslant j \leqslant 4)$$
$$\bar{b}_j = vb_j + v'b'_j + v''b''_j,$$

诸值 $H_1(\bar{a}_i, \bar{b}_j)$ 与 $H_2(\bar{a}_i, \bar{b}_j)$，将列表如下：

H_1	$\bar{a}_1$	$\bar{a}_2$	$\bar{a}_3$	$\bar{a}_4$
$\bar{b}_1$	$\bar{\gamma}^1_{11}$	$\bar{\beta}^1_{21}$	$\bar{\alpha}^1_{31}$	$\bar{\delta}^1_{41}$
$\bar{b}_2$	$\bar{\beta}^1_{12}$	$\bar{\alpha}^1_{22}$	$\bar{\delta}^1_{32}$	$\bar{\gamma}^1_{42}$
$\bar{b}_3$	$\bar{\alpha}^1_{13}$	$\bar{\delta}^1_{23}$	$\bar{\gamma}^1_{33}$	$\bar{\beta}^1_{43}$
$\bar{b}_4$	$\bar{\delta}^1_{14}$	$\bar{\gamma}^1_{24}$	$\bar{\beta}^1_{34}$	$\bar{\alpha}^1_{44}$

H_2	$\bar{a}_1$	$\bar{a}_2$	$\bar{a}_3$	$\bar{a}_4$
$\bar{b}_1$	$\bar{\beta}^2_{11}$	$\bar{\gamma}^2_{21}$	$\bar{\delta}^2_{31}$	$\bar{\alpha}^2_{41}$
$\bar{b}_2$	$\bar{\gamma}^2_{12}$	$\bar{\delta}^2_{22}$	$\bar{\alpha}^2_{32}$	$\bar{\beta}^2_{42}$
$\bar{b}_3$	$\bar{\delta}^2_{13}$	$\bar{\alpha}^2_{23}$	$\bar{\beta}^2_{33}$	$\bar{\gamma}^2_{43}$
$\bar{b}_4$	$\bar{\alpha}^2_{14}$	$\bar{\beta}^2_{24}$	$\bar{\gamma}^2_{34}$	$\bar{\delta}^2_{44}$

对任意按以上所选的数 (u, u', u'') 与 (v, v', v'')，在 $c \to 1$ 时显有 $H_1(\bar{a}_i, \bar{b}_j) \to H_1(a_i, b_j)$, $H_2(\bar{a}_i, \bar{b}_j) \to H_2(a_i, b_j)$. 因之我们可取 $c > 0$ 充分接近于 1，使诸值 $\bar{\alpha}, \bar{\beta}, \bar{\gamma}, \bar{\delta}$ 之间的大小关系与不等式 (1)—(4) 所示者相同，例如：

$$\bar{\delta}^k_{i_1,j_1} < \bar{\alpha}^k_{i_2,j_2} < \bar{\beta}^k_{i_3,j_3} < \bar{\gamma}^k_{i_4,j_4}, \quad (\bar{1})$$

$$2\bar{\delta}^k_{i_1,j_1} < \bar{\alpha}^k_{i_2,j_2} \quad (\bar{2})$$

以及 $(\bar{3})$, $(\bar{4})$ $k = 1, 2$, 而 $i_r, j_r = 1, 2, 3, 4$ 任意. 今设 P_1 与 P_2 各为闭多角形 $\bar{a}_1\bar{a}_2\bar{a}_3\bar{a}_4\bar{a}_1$ 与 $\bar{b}_1\bar{b}_2\bar{b}_3\bar{b}_4\bar{b}_1$. 空间 $P_1 \times P_2$ 在拓扑上是一环面，我们将表示为一正方形，其对边彼此恒同. 假设在 $P_1 \times P_2$ 上有平衡局势 (μ_1^*, μ_2^*)，我们将依照 $u, u', \cdots$ 的值区别几种不同情形来证明这不可能.

情形 I. $u > 0, v > 0$.

由于改变区域的选择，可见在这时对 P_1 上 μ_1^* 的某一邻域中的任一 μ_1，有

$$H_1(\mu_1^*, \mu_2^*) \geqslant H_1(\mu_1, \mu_2^*), \tag{5}$$

同样对 P_2 上 μ_2^* 的某一邻域中的任一 μ_2，有

$$H_2(\mu_1^*, \mu_2^*) \geqslant H_2(\mu_1^*, \mu_2). \tag{6}$$

由于 $(\bar{1})$—$(\bar{4})$ 诸不等式，可见为满足不等式 (5) 与 (6)，点 (μ_1^*, μ_2^*) 必须位于图中的粗黑线上 (图中 $\bar{\alpha}, \cdots, \bar{\delta}$ 各指 $H_1(\bar{a}_i, \bar{b}_j)$ 与 $H_2(\bar{a}_i, \bar{b}_j)$ 的值而为 $\bar{\alpha}^k_{ij}$ 等的简写). 因这些粗黑线互不相交，故这样的平衡局势不能存在.

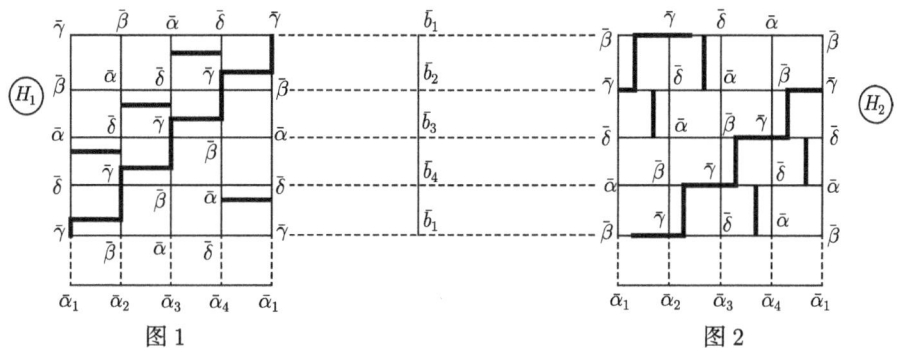

图 1　　　　　　　　　　　　图 2

情形 II. $u > 0, v = 0$.

这时 (5) 对于 P_1 上 μ_1^* 的某一邻域中的任意 μ_1 仍应满足如前. 至于 (6), 则只在 $\mu_2^* \neq \bar{b}_1, \bar{b}_2, \bar{b}_3$ 或 $\bar{b}_4$, 仍应对 P_2 上 μ_2^* 的某一邻域中的任意 μ_2 都满足. 因之 $P_1 \times P_2$ 上的平衡局势如果存在, 必须一方面位于图 1 中的粗黑线上, 另一面又须在图 2 中的粗黑线或水平线上. 这样的平衡局势只可能是

$$(\bar{a}_1, \bar{b}_1), (\bar{a}_2, \bar{b}_4), (\bar{a}_3, \bar{b}_3) \text{ 或 } (\bar{a}_4, \bar{b}_2).$$

但我们有

$$H_2(a_1, b_1) - H_2(a_1, b_1') = (1-c)(\beta - \delta) > 0,$$
$$H_2(a_1, b_1) - H_2(a_1, b_1'') = (1-c)(2\beta - \gamma - \alpha) > 0,$$
$$H_2(a_2, b_1) - H_2(a_2, b_1') = (1-c)(\gamma - \alpha),$$
$$H_2(a_2, b_1) - H_2(a_2, b_1'') = (1-c)(2\gamma - \beta - \delta),$$
$$H_2(a_3, b_1) - H_2(a_3, b_1') = (1-c)(\delta - \beta),$$
$$H_2(a_3, b_1) - H_2(a_3, b_1'') = (1-c)(2\delta - \alpha - \gamma),$$
$$H_2(a_4, b_1) - H_2(a_4, b_1') = (1-c)(\alpha - \gamma),$$
$$H_2(a_4, b_1) - H_2(a_4, b_1'') = (1-c)(2\alpha - \beta - \delta).$$

由此得 $c \to 1$ 时

$$\frac{1}{1-c}[H_2(\bar{a}_1, b_1) - H_2(\bar{a}_1, \bar{b}_1)] \to v'(\beta - \delta) + v''(2\beta - \gamma - \alpha) > 0.$$

因之只须 c 充分接近于 1, 即有

$$H_2(\bar{a}_1, b_1) > H_2(\bar{a}_1, \bar{b}_1).$$

因 $b_1 \in \tau_2(\bar{b}_1)$, 故上一不等式说明 $(\bar{a}_1, \bar{b}_1)$ 不能是平衡局势. 同样 $(\bar{a}_2, \bar{b}_4)$ 等也不能是平衡局势. 故在这一情形中没有平衡局势.

情形 III. $u=0, v>0$.

在这一情形中在 $P_1 \times P_2$ 上的平衡局势应位于图 2 中的粗黑线上又位于图 1 中粗黑线或垂直线上. 唯一的可能性是

$$(\bar{a}_1, \bar{b}_2), \quad (\bar{a}_2, \bar{b}_1), \quad (\bar{a}_3, \bar{b}_4) \quad 或 \quad (\bar{a}_4, \bar{b}_3).$$

像情形 II 那样, 这些都是不可能的.

情形 IV. $u=0, v=0$.

如前, 唯一的可能性是以下 16 个点:

$$(\bar{a}_i, \bar{b}_j), \quad i,j = 1,2,3,4.$$

但点

$$(\bar{a}_1, \bar{b}_1), \quad (\bar{a}_2, \bar{b}_4), \quad (\bar{a}_3, \bar{b}_3), \quad (\bar{a}_4, \bar{b}_2)$$

在情形 II 中已知其不可能, 而点

$$(\bar{a}_1, \bar{b}_2), \quad (\bar{a}_2, \bar{b}_1), \quad (\bar{a}_3, \bar{b}_4), \quad (\bar{a}_4, \bar{b}_3)$$

在情形 III 中又已知其不可能. 其余需要试验的点是

$$(\bar{a}_1, \bar{b}_3), \quad (\bar{a}_2, \bar{b}_2), \quad (\bar{a}_3, \bar{b}_1), \quad (\bar{a}_4, \bar{b}_4),$$
$$(\bar{a}_1, \bar{b}_4), \quad (\bar{a}_2, \bar{b}_3), \quad (\bar{a}_3, \bar{b}_2), \quad (\bar{a}_4, \bar{b}_1).$$

对于点 $(\bar{a}_1, \bar{b}_3)$ 置

$$b_3^* = cb_3 + (1-c)b_2.$$

于是

$$H_2(a_1, b_3^*) - H_2(a_1, b_3') = (1-c)(\gamma - \beta) > 0,$$
$$H_2(a_1, b_3^*) - H_2(a_1, b_3'') = (1-c)(2\delta - \alpha) > 0$$

等等. 由此得 c 充分接近于 1 时

$$H_2(\bar{a}_1, b_3^*) - H_2(\bar{a}_1, \bar{b}_3) > 0.$$

因 $b_3^* \in \tau_2(\bar{b}_3)$, 故 $(\bar{a}_1, \bar{b}_3)$ 不能是一个平衡局势. 诸点 $(\bar{a}_2, \bar{b}_2), (\bar{a}_3, \bar{b}_1)$ 与 $(\bar{a}_4, \bar{b}_4)$ 也是如此.

对于点 $(\bar{a}_1, \bar{b}_4)$ 置

$$a_1^* = ca_1 + (1-c)a_2.$$

则
$$H_1(a_1^*, b_4) - H_1(a_1', b_4) = (1-c)(\gamma - \beta) > 0,$$
$$H_1(a_1^*, b_4) - H_1(a_1'', b_4) = (1-c)(2\delta - \alpha) > 0$$

等等. 由此得 c 充分接近于 1 时

$$H_1(a_1^*, \bar{b}_4) - H_1(\bar{a}_1, \bar{b}_4) > 0.$$

因 $a_1^* \in \tau_1(\bar{a}_1)$, 故 $(\bar{a}_1, \bar{b}_4)$ 不能是平衡局势. 诸点 $(\bar{a}_2, \bar{b}_3)$, $(\bar{a}_3, \bar{b}_2)$ 与 $(\bar{a}_4, \bar{b}_1)$ 亦然.

综合上述可见只须 c 充分接近于 1, 我们的活动受限制对策即无平衡局势, 虽然每个对策者只拥有有限个纯正策略.

参考文献

[1] E. G. Begle. The Vietoris mapping theorem for bicompact spaces. *Annals of Math.*, 1950, 51: 534-543.

[2] Dunford-Schwartz. Linear operators, part I, General theory. New York, 1958.

[3] I. L. Glicksberg. A further generalization of the Kakutani fixed point theorem, with application to Nash equilibrium points. *Proc. Amer. Math. Soc.*, 1952, 3: 170-174.

[4] J. Leray. Sur la forme des espaces topologiques et sur les points fixes des représentations. *J. de Math.*, 1945, 24: 95-167.

[5] J. Nash. Non-coorperative games. *Annals of Math.*, 1951, 54: 286-295.

Essential Equilibrium Points of n-Person Non-cooperative Games*

§1. Introduction

An n-person non-cooperative game is completely determined by its pay-off functions if its sets of strategies are fixed once for all. Therefore, it is clear that the existence and the characters of the equilibrium points of a game depend on the evaluations of its pay-off functions. It is possible that equilibrium points which have been determined by an inaccurate evaluation of pay-off functions are not "true" ones. We shall introduce in this paper the concept, analogous to the notion of the usual stability, of essential equilibrium points for n-person non-cooperative finite games, which, as an equilibrium point, fails to "disappear" by negligible error of evaluation of pay-off functions. Moreover, we shall prove that any game may be approximated arbitrarily closely by a game whose equilibrium points are all essential, and that any game having only a finite number of equilibrium points has at least one essential equilibrium point.

§2. The notion of essential equilibrium points

Let
$$\Gamma = \langle I, \{S_i\}_{i \in I}, \{H_i\}_{i \in I} \rangle$$
be an n-person non-cooperative game for which $I = \{1, \cdots, n\}$ denotes the set of players, and each player possesses only a finite set of pure strategies
$$S_i = \{\pi^i_{a_i} | a_i \in M_i = \{1, \cdots, m_i\}\}, \quad (i \in I).$$
We shall call $S = S_1 \times \cdots \times S_n$ the set of pure situations of Γ and shall write $M = M_1 \times \cdots \times M_n$. Under the pure situation $\pi_a = (\pi^1_{a_1}, \cdots, \pi^n_{a_n}) \in S$, the pay-off

---

* 本文原载 *Scientia Sinica*, 1962, 11. 作者: Wu Wen-Tsün (吴文俊) and Jiang Jia-He (江嘉禾).

of i-th player is given by

$$H_i(\pi_a) = a_a^i \quad (i \in I, a = (a_1, \cdots, a_n) \in M). \tag{1}$$

In all our following considerations I and S_i will be kept fixed. The game Γ is then completely determined by the set of numbers $a = (a_a^i)_{i \in I, a \in M}$, which will be called the *determining set of* Γ. For two games Γ_a and Γ_b with determining sets $a = (a_a^i)$ and $b = (b_a^i)$ we shall call

$$\mathscr{D}(\Gamma_a, \Gamma_b) = \sum_{\substack{i \in I \\ a \in M}} |a_a^i - b_a^i| \tag{2}$$

the *distance* between Γ_a and Γ_b. The set $\mathscr{G}$ of all such games becomes then a complete metric space with $\mathscr{D}$ as a metric function.

Let

$$\Gamma^* = \langle I, \{S_i^*\}_{i \in I}, \{H_i^*\}_{i \in I} \rangle$$

be the natural extension of Γ with determining set $a = (a_a^i)$ in which

$$S_i^* = \left\{ x^i = \sum_{a_i \in M_i} x_{a_i}^i \pi_{a_i}^i \,\middle|\, x_{a_i}^i \geq 0, \sum_{a_i \in M_i} x_{a_i}^i = 1 \right\} \quad (i \in I)$$

is the set of mixed strategies of player "i", and $S^* = S_1^* \times \cdots \times S_n^*$ will be called the space of situations of Γ^*. Under the situation $x = (x^1, \cdots, x^n) \in S^*$, the pay-off of i-th player is given by

$$H_i^*(x) = H_i^*(x^1, \cdots, x^n) = \sum_{a=(a_1,\cdots,a_n) \in M} a_a^i x_{a_1}^1 \cdots x_{a_n}^n, \quad (i \in I). \tag{3}$$

This is the corresponding extension of the pay-off function of player "i".

According to J. Nash, a situation $x = (x^1, \cdots, x^n) \in S^*$ is called an *equilibrium point* of the game Γ if for each i,

$$H_i^*(x) = \sup_{y^i \in S_i^*} H_i^*(x|y^i), \tag{4}$$

in which $(x|y^i) = (x^1, \cdots, x^{i-1}, y^i, x^{i+1}, \cdots, x^n)$ denotes the situation deduced from x in replacing $x^i \in S_i^*$ by $y^i \in S_i^*$.

In order to introduce the concept of essential equilibrium points, let us call

$$d(x, y) = \sum_{\substack{i \in I \\ a_i \in M_i}} |x_{a_i}^i - y_{a_i}^i| \tag{5}$$

the *distance* between two situations $x = (x^1, \cdots, x^n) \in S^*$ and $y = (y^1, \cdots, y^n) \in S^*$ in *which* $x^i = \sum_{a_i} x^i_{a_i} \pi^i_{a_i} \in S^*_i$ and $y^i = \sum_{a_i} y^i_{a_i} \pi^i_{a_i} \in S^*_i (i \in I)$. The space S^* becomes then a compact metric space with d as a metric function. We shall introduce then the following conception:

Definition. An equilibrium point $x \in S^*$ of the game $\Gamma \in \mathscr{G}$ will be called an *essential equilibrium point* of Γ if for every $\varepsilon > 0$, there is a $\delta > 0$ such that for any game $\tilde{\Gamma} \in \mathscr{G}$ with $\mathscr{D}(\Gamma, \tilde{\Gamma}) < \delta$, there exists at least one equilibrium point $\tilde{x}$ of $\tilde{\Gamma}$ with $d(x, \tilde{x}) < \varepsilon$. A game $\Gamma \in \mathscr{G}$ will be called an *essential game* if all its equilibrium points are essential.

Easy examples show that a game, e. g., the game with determining set $(a^i_a = 0)_{i \in I, a \in M}$, may have no essential equilibrium point at all, though equilibrium points necessarily exist, according to the fundamental theorem of Nash.

Our purpose is to prove the following theorems:

Theorem A *Any game may be approximated arbitrarily near to it by an essential game. More precisely, for any game $\Gamma \in \mathscr{G}$ and any $\varepsilon > 0$, there is an essential game $\tilde{\Gamma} \in \mathscr{G}$ such that $\mathscr{D}(\Gamma, \tilde{\Gamma}) < \varepsilon$.*

Theorem B *Any game having only a finite number of equilibrium points has among them at least one essential equilibrium point.*

§ 3. The Nash mapping

For the compact metric space S^* described in §2, we shall denote by $C(S^*)$ the space of all continuous mappings of S^* into itself. For any $f \in C(S^*)$ and any $g \in C(S^*)$, we define the *distance* between f and g by

$$\rho(f, g) = \sup_{x \in S^*} d(f(x), g(x)). \tag{6}$$

Then $C(S^*)$ becomes a complete metric space with ρ as a metric function.

Given a game $\Gamma = \langle I, \{S_i\}_{i \in I}, \{H_i\}_{i \in I} \rangle$ with natural extension $\Gamma^* = \langle I, \{S^*_i\}_{i \in I}, \{H^*_i\}_{i \in I} \rangle$. Let us put

$$\varphi^i_{\beta_i}(x) = \max\{0, H^*_i(x|\pi^i_{\beta_i}) - H^*_i(x)\}, \tag{7}$$

for any $x \in S^*, i \in I$ and $\beta_i \in M_i$. The mapping

$$f_r : S^* \to S^*,$$

defined by
$$f_r(x) = \bar{x} = (\bar{x}^1, \cdots, \bar{x}^n),$$
in which
$$\bar{x}^i = \frac{x^i + \sum_{\beta_i \in M_i} \varphi^i_{\beta_i}(x) \pi^i_{\beta_i}}{1 + \sum_{\beta_i \in M_i} \varphi^i_{\beta_i}(x)}, \quad (i \in I) \tag{8}$$

or, in details,
$$\bar{x}^i_{\alpha_i} = \frac{x^i_{\alpha_i} + \varphi^i_{\alpha_i}(x)}{1 + \sum_{\beta_i \in M_i} \varphi^i_{\beta_i}(x)}, \quad (i \in I, \alpha_i \in M_i) \tag{9}$$

will be called the *Nash mapping* of the game Γ. It is clear that $f_r \in C(S^*)$. The following fact plays then a central role in the theory of Nash:

Lemma 1 $x \in S^*$ *is an equilibrium point of* Γ *if and only if x is a fixed point of f_r.*

In view of this Lemma our method of proving Theorem A and Theorem B consists now in the study of interrelations between games Γ and their Nash mappings f_r. To begin with, we shall introduce the following.

Definition. Two games Γ_a, Γ_b with determining sets $a = (a^i_a)$ and $b = (b^i_a)$ will be said to be *isomorphic*, in symbol:
$$\Gamma_a \approx \Gamma_b,$$
if for each $i \in I$ and each $a = (a_1, \cdots, a_n) \in M$, the number $a^i_a - b^i_a$ is independent of a_i:
$$a^i_a = b^i_a + \mu^i_{(a||i)}, \tag{10}$$
in which $(a||i)$ denotes $(a_1, \cdots, a_{i-1}, a_{i+1}, \cdots, a_n)$, and $\mu^i_{(a||i)}$ denotes a number, independent of a_i. They will be said to be *equivalent*, in symbol:
$$\Gamma_a \sim \Gamma_b,$$
if there exist numbers $\lambda^i > 0 (i \in I)$ and numbers $\mu^i_{(a||i)} (i \in I, a \in M)$, independent of $a_i \in M_i$ such that
$$a^i_a = \lambda^i b^i_a + \mu^i_{(a||i)} \tag{11}$$
for each $i \in I$ and each $a \in M$.

Lemma 2 If $\Gamma_a \sim \Gamma_b$, then Γ_a and Γ_b have not only the same set of equilibrium points, but also the same set of essential equilibrium points. Therefore, if one of them is an essential game, then such will the other be.

Proof. For the pay-off functions $H_{i,a}^*$ and $H_{i,b}^*$ of the extended games Γ_a^* and Γ_b^*, we have by (3) and (11),

$$H_{i,a}^*(x) = \sum_a a_a^i x_{a_1}^1 \cdots x_{a_n}^n$$
$$= \sum_a (\lambda^i b_a^i + \mu_{(a||i)}^i) x_{a_1}^1 \cdots x_{a_n}^n$$
$$= \lambda^i \sum_a b_a^i x_{a_1}^1 \cdots x_{a_n}^n + \sum_{(a||i)} \mu_{(a||i)}^i x_{a_1}^1 \cdots x_{a_{i-1}}^{i-1} x_{a_{i+1}}^{i+1} \cdots x_{a_n}^n$$
$$= \lambda^i H_{i,b}^*(x) + \sum_{(a||i)} \mu_{(a||i)}^i x_{a_1}^1 \cdots x_{a_{i-1}}^{i-1} x_{a_{i+1}}^{i+1} \cdots x_{a_n}^n,$$
$$H_{i,a}^*(x|y^i) = \lambda^i H_{i,b}^*(x|y^i) + \sum_{(a||i)} \mu_{(a||i)}^i x_{a_1}^1 \cdots x_{a_{i-1}}^{i-1} x_{a_{i+1}}^{i+1} \cdots x_{a_n}^n.$$

Hence
$$H_{i,a}^*(x) - H_{i,a}^*(x|y^i) = \lambda^i (H_{i,b}^*(x) - H_{i,b}^*(x|y^i)).$$

Since $\lambda^i > 0$, it is clear that
$$H_{i,a}^*(x) - H_{i,a}^*(x|y^i) \text{ and } H_{i,b}^*(x) - H_{i,b}^*(x|y^i),$$
have the same sign, i. e., Γ_a and Γ_b have the same set of equilibrium points.

Now let x be an essential equilibrium point of Γ_a: for every $\varepsilon > 0$, there is a $\delta > 0$ such that, for any game $\Gamma \in \mathscr{G}$ with $\mathscr{D}(\Gamma_a, \Gamma) < \delta$, there exists at least one equilibrium point $\widetilde{x}$ of Γ with $d(x, \widetilde{x}) < \varepsilon$. We shall prove that x, as an equilibrium point of Γ_b, is also an essential equilibrium point of Γ_b. It is clear that the equivalence relation (11) between Γ_a and Γ_b is a topological transformation of the space $\mathscr{G}$ onto itself, the numbers λ^i and $\mu_{(a||i)}^i$ being kept fixed. Hence, there is a $\delta_1 > 0$ such that $\mathscr{D}(\Gamma_a, \Gamma) < \delta$ whenever $\mathscr{D}(\Gamma_b, \widetilde{\Gamma}) < \delta_1$, where Γ denotes the game deduced from $\widetilde{\Gamma}$ by transformation (11). Hence, there exists one equilibrium point $\widetilde{x}$ of Γ with $d(x, \widetilde{x}) < \varepsilon$, and $\widetilde{x}$ is also an equilibrium point of $\widetilde{\Gamma}$ since $\Gamma \sim \widehat{\Gamma}$. It follows that, for any game $\widetilde{\Gamma} \in \mathscr{G}$ with $\mathscr{D}(\Gamma_b, \widetilde{\Gamma}) < \delta_1$, there exists at least one equilibrium point $\widetilde{x}$ of $\widetilde{\Gamma}$ with $d(x, \widetilde{x}) < \varepsilon$. This proves that x is an essential equilibrium point of Γ_b. Similarly, x will be an essential equilibrium point of Γ_a if it is an essential equilibrium point of Γ_b. Therefore equivalent games have the same set of essential equilibrium points, and the proof is complete.

Lemma 3 *Two games Γ_a and Γ_b have the same Nash mapping if and only if $\Gamma_a \approx \Gamma_b$.*

Proof. *Necessity.* Let $f_{\Gamma_a} = f_{\Gamma_b}$, then for any pure situation $\pi_a = (\pi_{a_1}^1, \cdots, \pi_{a_n}^n) \in S$, we have $f_{\Gamma_a}(\pi_a) = f_{\Gamma_b}(\pi_a)$, so that

$$\frac{\pi_{a_i}^i + \sum_{\beta_i} \varphi_{\beta_i}^{i,a}(\pi_a)\pi_{\beta_i}^i}{1 + \sum_{\beta_i} \varphi_{\beta_i}^{i,a}(\pi_a)} = \frac{\pi_{a_i}^i + \sum_{\beta_i} \varphi_{\beta_i}^{i,b}(\pi_a)\pi_{\beta_i}^i}{1 + \sum_{\beta_i} \varphi_{\beta_i}^{i,b}(\pi_a)},$$

in which $\varphi_{\beta_i}^{i,a}, \varphi_{\beta_i}^{1,b}$ are the $\varphi_{\beta_i}^i$ corresponding to Γ_a and Γ_b respectively. As

$$\varphi_{a_i}^{i,a}(\pi_a) = \varphi_{a_i}^{i,b}(\pi_a) = 0$$

by definition of $\varphi_{\beta_i}^i$, we get by comparing coefficient of $\pi_{a_i}^i$

$$\sum_{\beta_i} \varphi_{\beta_i}^{i,a}(\pi_a) = \sum_{\beta_i} \varphi_{\beta_i}^{i,b}(\pi_a),$$

whence

$$\sum_{\beta_i} \varphi_{\beta_i}^{i,a}(\pi_a)\pi_{\beta_i}^i = \sum_{\beta_i} \varphi_{\beta_i}^{i,b}(\pi_a)\pi_{\beta_i}^i.$$

By comparing coefficient of $\pi_{\beta_i}^i$, we get then for any $a \in M$ and any $\beta_i \in M_i$,

$$\varphi_{\beta_i}^{i,a}(\pi_a) = \varphi_{\beta_i}^{i,b}(\pi_a),$$

i. e.,

$$\max\{0, a_{(a|\beta_i)}^i - a_a^i\} = \max\{0, b_{(a|\beta_i)}^i - b_a^i\}.$$

It follows

$$a_{(a|\beta_i)}^i - a_a^i = b_{(a|\beta_i)}^i - b_a^i$$

by the arbitrariness of a and β_i, or

$$a_{(a|\beta_i)}^i - b_{(a|\beta_i)}^i = a_a^i - b_a^i = \mu_{(a||i)}^i$$

is independent of a_i in a, i. e., $\Gamma_a \approx \Gamma_b$.

Sufficiency. Let $\Gamma_a \approx \Gamma_b$, so that for each $i \in I$ and each $a \in M$, we have

$$a_a^i = b_a^i + \mu_{(a||i)}^i,$$

in which $\mu^i_{(a||i)}$ is independent of a_i, then

$$H^*_{i,x}(x) = \sum_a a^i_a x^1_{a_1} \cdots x^n_{a_n}$$
$$= \sum_a b^i_a x^1_{a_1} \cdots x^n_{a_n} + \sum_{(a||i)} \mu^i_{(a||i)} x^1_{a_1} \cdots x^{i-1}_{a_{i-1}} x^{i+1}_{a_{i+1}} \cdots x^n_{a_n}$$
$$= H^*_{i,b}(x) + \sum_{(a||i)} \mu^i_{(a||i)} x^1_{a_1} \cdots x^{i-1}_{a_{i-1}} x^{i+1}_{a_{i+1}} \cdots x^n_{a_n},$$
$$H^*_{i,a}(x|y^i) = H^*_{i,b}(x|y^i) + \sum_{(a||i)} \mu^i_{(a||i)} x^1_{a_1} \cdots x^{i-1}_{a_{i-1}} x^{i+1}_{a_{i+1}} \cdots x^n_{a_n}.$$

Hence
$$H^*_{i,x}(x|y^i) - H^*_{i,a}(x) = H^*_{i,x}(x|y^i) - H^*_{i,b}(x)$$

and
$$\varphi^{i,a}_{\beta_i}(x) = \varphi^{i,b}_{\beta_i}(x), \quad (x \in S^*)$$

for each $i \in I$ and each $\beta_i \in M_i$. Thus $\Gamma_a \approx \Gamma_b$ implies $f_{\Gamma_a} = f_{\Gamma_b}$, and the proof is complete.

Since corresponding to every game $\Gamma \in \mathscr{G}$, there is a continuous mapping $f_\Gamma \in C(S^*)$, we obtain the mapping

$$h : \mathscr{G} \longrightarrow C(S^*),$$

with thus $h(\Gamma) = f_\Gamma$.

Lemma 4 $h : \mathscr{G} \longrightarrow C(S^*)$ *is a continuous mapping.*

Proof. We shall prove that h is continuous at any $\Gamma_a \in \mathscr{G}$. Suppose that for the determining set $a = (a^i_a)$ of Γ_a,

$$|a^i_a| \leqslant N \quad (i \in I, a \in M).$$

For any $\varepsilon > 0$, let

$$\delta < \frac{\varepsilon}{4nm(1 + m + 4mN)}$$

in which $m = m_1 \cdots m_n$. Suppose that $\mathscr{D}(\Gamma_a, \Gamma_b) < \delta$, we shall prove that $\rho(f_{\Gamma_a}, f_{\Gamma_b}) < \varepsilon$. Let $b = (b^i_a)$ be the determining set of Γ_b, then for any $i \in I$ and $a \in M$,

$$|a^i_a - b^i_a| < \delta.$$

Hence for any $x \in S^*$, we have

$$|H^*_{i,a}(x) - H^*_{i,b}(x)| < \delta, \quad (i \in I).$$

Therefore,
$$|[H^*_{i,a}(x|\pi^i_{\beta_i}) - H^*_{i,a}(x)] - [H^*_{i,b}(x|\pi^i_{\beta_i}) - H^*_{i,b}(x)]|$$
$$\leqslant |H^*_{i,a}(x|\pi^i_{\beta_i}) - H^*_{i,b}(x|\pi^i_{\beta_i})| + |H^*_{i,a}(x) - H^*_{i,b}(x)| < 2\delta$$

for any $i \in I$, $\pi^i_{\beta_i} \in S_i$ and $x \in S^*$. By the definition of $\varphi^i_{\beta_i}(x)$, we obtain

$$\left|\varphi^{i,a}_{\beta_i}(x) - \varphi^{i,b}_{\beta_i}(x)\right| < 2\delta,$$

whence

$$\left|\sum_{\beta_i} \varphi^{i,a}_{\beta_i}(x) - \sum_{\beta_i} \varphi^{i,b}_{\beta_i}(x)\right| < 2m\delta.$$

Moreover, it is clear that

$$\left|H^*_{i,a}(x|\pi^i_{\beta_i}) - H^*_{i,a}(x)\right| \leqslant 2N,$$
$$\varphi^{i,a}_{\beta_i}(x) \leqslant 2N,$$
$$\sum_{\beta_i} \varphi^{i,a}_{\beta_i}(x) \leqslant 2mN.$$

Now we shall evaluate $d(f_{\Gamma_a}(x), f_{\Gamma_b}(x))$ for any $x = (x^1, \cdots, x^n) \in S^*$ with $x^i = \sum_{a_i} x^i_{a_i} \pi^i_{a_i}$. We have

$$\left|\frac{x^i_{a_i} + \varphi^{i,a}_{a_i}(x)}{1 + \sum_{\beta_i} \varphi^{i,a}_{\beta_i}(x)} - \frac{x^i_{a_i} + \varphi^{i,b}_{a_i}(x)}{1 + \sum_{\beta_i} \varphi^{i,b}_{\beta_i}(x)}\right|$$

$$\leqslant \left|\left[1 + \sum_{\beta_i} \varphi^{i,b}_{\beta_i}(x)\right]\left[x^i_{a_i} + \varphi^{i,a}_{a_i}(x)\right]\right.$$

$$\left. - \left[1 + \sum_{\beta_i} \varphi^{i,a}_{\beta_i}(x)\right]\left[x^i_{a_i} + \varphi^{i,b}_{a_i}(x)\right]\right|$$

$$\leqslant |\varphi^{i,a}_{a_i}(x) - \varphi^{i,b}_{a_i}(x)| + x^i_{a_i}\left|\sum_{\beta_i} \varphi^{i,b}_{\beta_i}(x) - \sum_{\beta_i} \varphi^{i,a}_{\beta_i}(x)\right|$$

$$+ \varphi^{i,a}_{a_i}(x)\left|\sum_{\beta_i} \varphi^{i,b}_{\beta_i}(x) - \sum_{\beta_i} \varphi^{i,a}_{\beta_i}(x)\right|$$

$$+ \left|\varphi^{i,a}_{a_i}(x) - \varphi^{i,b}_{a_i}(x)\right|\sum_{\beta_i} \varphi^{i,a}_{\beta_i}(x)$$

$$\leqslant 2\delta + 2m\delta + 2N \cdot 2m\delta + 2mN \cdot 2\delta$$
$$= 2(1 + m + 4mN)\delta.$$

Thus, for any $x \in S^*$, we have

$$d(f_{\Gamma_a}(x), f_{\Gamma_b}(x)) = \sum_{i,a_i} \left| \frac{x^i_{a_i} + \varphi^{i,a}_{a_i}(x)}{1 + \sum_{\beta_i} \varphi^{i,a}_{\beta_i}(x)} - \frac{x^i_{a_i} + \varphi^{i,b}_{a_i}(x)}{1 + \sum_{\beta_i} \varphi^{i,b}_{\beta_i}(x)} \right|$$

$$\leqslant 2nm(1 + m + 4mN)\delta < \varepsilon/2.$$

Therefore, we get

$$\rho(f_{\Gamma_a}, f_{\Gamma_b}) = \sup_{x \in S^*} d(f_{\Gamma_a}(x), f_{\Gamma_b}(x)) < \varepsilon.$$

This proves the continuity of h at Γ_a, and the proof is complete.

§4. Normalization

Definition. A game Γ with determining set $a = (a^i_a)$ will be said to be *normalized* if for each $i \in I$ and each $a \in M$, we have

(i) $P^i_{(a||i)}(\Gamma) = \sum_{\beta_i \in M_i} a^i_{(\alpha|\beta_i)} = 0,$

(ii) $N^i(\Gamma) = \sum_{\substack{a \in M \\ \beta_i, \gamma_i \in M_i}} |a^i_{(a|\beta_i)} - a^i_{(a|\gamma_i)}| = $ either 0 or 1.

Lemma 5 *For normalized games, $\Gamma_a \sim \Gamma_b$ if and only if $\Gamma_a = \Gamma_b$.*

Proof. Let $a = (a^i_a)$ and $b = (b^i_a)$ be determining sets of two normalized games Γ_a and Γ_b, respectively. Suppose that $\Gamma_a \sim \Gamma_b$, so that there are $\lambda^i > 0$ and $\mu^i_{(a||i)}$ independent of $\alpha^i \in M_i$ such that

$$a^i_a = \lambda^i b^i_a + \mu^i_{(a||i)}$$

for any $i \in I$ and $\alpha \in M$. Hence, we have

$$N^i(\Gamma_a) = \sum_{a, \beta_i, \gamma_i} |a^i_{(a|\beta_i)} - a^i_{(a|\gamma_i)}|$$
$$= \lambda^i \sum_{a, \beta_i, \gamma_i} |b^i_{(a|\beta_i)} - b^i_{(a|\gamma_i)}| = \lambda^i N^i(\Gamma_b).$$

Therefore, $\lambda^i = 1$ in case $N^i(\Gamma_a) = N^i(\Gamma_b) = 1$. Moreover in such case we have

$$P^i_{(a||i)}(\Gamma_a) = \sum_{\beta_i} a^i_{(a|\beta_i)} = \sum_{\beta_i} b^i_{(a|\beta_i)} + \sum_{\beta_i} \mu^i_{(a||i)}$$
$$= P^i_{(a||i)}(\Gamma_b) + m_i \mu^i_{(a||i)}.$$

As $P^i_{(a||i)}(\Gamma_a) = P^i_{(a||i)}(\Gamma_b) = 0$, we obtain $\mu^i_{(a||i)} = 0$, i. e., $a^i_a = b^i_a$. In the contrary case, we have $N^i(\Gamma_a) = N^i(\Gamma_b) = 0$ so that

$a^i_{(a|\beta_i)}$ is independent of β_i and $= A^i_{(a||i)}$,

$b^i_{(a|\beta_i)}$ is independent of β_i and $= B^i_{(a||i)}$.

As

$$P^i_{(a||i)}(\Gamma_a) = \sum_{\beta_i} a^i_{(a|\beta_i)} = m_i A^i_{(a||i)} = 0,$$

and

$$P^i_{(a||i)}(\Gamma_b) = \sum_{\beta_i} b^i_{(a|\beta_i)} = m_i B^i_{(a||i)} = 0,$$

we have then $a^i_a = b^i_a (= 0)$ again for any $a \in M$. Therefore, this proves that for any $i \in I$ and any $\alpha \in M$, we have $a^i_a = b^i_a$, i. e., $\Gamma_a = \Gamma_b$, and the proof is complete.

Lemma 6 *Any game is equivalent to one and only one normalized game.*

Proof. For any game Γ_a with determining set $a = (a^i_a)$, let us set

$$\mu^i_{(a||i)} = -\frac{1}{m_i} \sum_{\beta_i \in M_i} a^i_{(a|\beta_i)}, \tag{12}$$

$$\lambda^i = \begin{cases} 1, & \text{if } N^i(\Gamma_a) = 0, \\ 1/N^i(\Gamma_a), & \text{if } N^i(\Gamma_a) \neq 0, \end{cases} \tag{13}$$

$$b^i_a = \lambda^i(a^i_a + \mu^i_{(a||i)}).$$

Then the game Γ_b with determining set $b = (b^i_a)$ is evidently equivalent to Γ_a and is a normalized game. Since

$$P^i_{(a||i)}(\Gamma_b) = \sum_{\beta_i} b^i_{(a|\beta_i)} = \lambda^i \sum_{\beta_i} a^i_{(a|\beta_i)} + m_i \lambda^i \mu^i_{(a||i)}$$

$$= \lambda^i P^i_{(a||i)}(\Gamma_a) - \lambda^i P^i_{(a||i)}(\Gamma_a) = 0,$$

$$N^i(\Gamma_b) = \sum_{a, \beta_i, \gamma_i} |b^i_{(a|\beta_i)} - b^i_{(a|\gamma_i)}|$$

$$= \lambda^i \sum_{a, \beta_i, \gamma_i} |a^i_{(a|\beta_i)} - a^i_{(a|\gamma_i)}|$$

$$= \lambda^i N(\Gamma_a) = \text{either } 0 \text{ or } 1.$$

Thus Γ_a is equivalent to the normalized game Γ_b. By Lemma 5, such Γ_b is uniquely determined, and the proof is complete.

The normalized game Γ_b in Lemma 6 will be called the *nor-malization* of game Γ_a. Thus Lemma 6 asserts that any game is equivalent to its unique normalization. Now let $\mathscr{G}_0 \subset \mathscr{G}$ be the sub-space of all normalized games $\Gamma \in \mathscr{G}$.

Lemma 7 $\mathscr{G}_0$ *is a compact metric space.*

Proof. We shall prove that any sequence $\{\Gamma_m\}$ of normalized games contains a subsequence convergent in $\mathscr{G}_0$. In fact, for any normalized game Γ with determining set $a = (a_\alpha^i)$, as $N^i(\Gamma) = 1$, we have

$$|a_\alpha^i - a_{(\alpha\,|\,\beta_i)}^i| \leqslant 1$$

for each $i \in I$, $\alpha \in M$ and $\beta_i \in M_i$. Hence, as $P_{(\alpha\,||\,i)}^i = 0$, we have

$$|a_\alpha^i| = \frac{1}{m_i}\left|\sum_{\beta_i}(a_\alpha^i - a_{(\alpha\,|\,\beta_i)}^i)\right| \leqslant 1.$$

Now let $a_m = ({}^m a_\alpha^i)$ be the determining sets of the normalized games Γ_m ($m = 1, 2, \cdots$). Then, for any fixed $i \in I$ and $\alpha \in M$, the sequence $\{{}^m a_\alpha^i\}$ of real numbers is bounded. It follows that there exists a subsequence $\{\Gamma_{m_v}\}$ of sequence $\{\Gamma_m\}$ such that for any $i \in I$ and $\alpha \in M$, the sequence $\{{}^{m_v} a_\alpha^i\}$ of numbers is convergent. Let a_α^i be the limit of sequence $\{m_v a_\alpha^i\}$, then the subsequence $\{\Gamma_{m_v}\}$ is convergent to the game Γ with determining set $a = (a_\alpha^i)$. It is clear that

$$P_{(\alpha\,||\,i)}^i(\Gamma_{m_v}) \to P_{(\alpha\,||\,i)}^i(\Gamma), \quad N^i(\Gamma_{m_v}) \to N^i(\Gamma).$$

As $P_{(\alpha\,||\,i)}^i(\Gamma_{m_v}) = 0$ and $N^i(\Gamma_{m_v})$ = either 0 or 1, the limit $P_{(\alpha\,||\,i)}^i(\Gamma) = 0$ and $N^i(\Gamma)$ = either 0 or 1, i. e., Γ is also a normalized game, and the proof is complete.

§5. Proof of the main theorems

For the continuous mapping $h : \mathscr{G} \to C(S^*)$, let us denote by $C_0 \subset C(S^*)$ the subspace consisting of all Nash mappings corresponding to normalized games: $C_0 = h(\mathscr{G}_0)$. By Lemma 3 and Lemma 5, $h : \mathscr{G}_0 \to C_0$ is then a one-one continuous mapping. As $\mathscr{G}_0$ is compact, h is then a homoeomorphism of $\mathscr{G}_0$ onto C_0. Therefore, C_0 is also a compact metric space and in particular a complete metric space.

A fixed point $x \in S^*$ of a mapping $f \in C_0$ will be said to be *essential* with respect to C_0 if for every $\varepsilon > 0$, there is a $\delta > 0$ such that for any $g \in C_0$ with $\rho(f, g) < \delta$, there is a fixed point y of g with $d(x, y) < \varepsilon$. A mapping $f \in C_0$ will be said to be *essential* in C_0 if every fixed point of f is an essential fixed point of f

with respect to C_0. We are now in need of the following lemma which is due to M. K. Fort, Jr.[2] and has been recently extended to multivalued mappings[3].

Lemma 8[1] For every $f \in C_0$ and every $\varepsilon > 0$, there is a mapping $g \in C_0$ essential in C_0 such that $\rho(f, g) < \varepsilon$.

Now let $\mathscr{G}^* \subset \mathscr{G}_0$ be the subspace consisting of all normalized games $\Gamma \in \mathscr{G}_0$ for which all $N^i(\Gamma)$ are equal to 1.

Lemma 9 For $\Gamma \in \mathscr{G}^*, x \in S^*$ is an essential equilibrium point of Γ if and only if x is an essential fixed point of f_r with respect to C_0. Therefore, $\Gamma \in \mathscr{G}^*$ is an essential game if and only if f_r is a mapping essential in C_0.

Proof. Since $\mathscr{G}_0$ and C_0 are homoeomorphic, the necessity is evident. Now we shall prove the sufficiency as follows. Suppose that $x \in S^*$ is an essential fixed point of f_r with respect to C_0: for every $\varepsilon > 0$, there is a $\delta > 0$ such that for any $f_{\tilde{r}} \in C_0$ with $\rho(f_r, f_{\tilde{r}}) < \delta$, there is a fixed point $\tilde{x}$ of $f_{\tilde{r}}$ with $d(x, \tilde{x}) < \varepsilon$. Since $\mathscr{G}_0$ and C_0 are homocomorphic, there is a $\delta_1 > 0$ such that for any $\tilde{\Gamma} \in \mathscr{G}_0$ with $\mathscr{D}(\Gamma, \tilde{\Gamma}) < \delta_1$, we have $\rho(f_r, f_{\tilde{r}}) < \delta$. Thus, for the equilibrium point x of Γ and every $\varepsilon > 0$, there is a $\delta_1 > 0$ such that for any $\tilde{\Gamma} \in \mathscr{G}_0$ with $\mathscr{D}(\Gamma, \tilde{\Gamma}) < \delta_1$, there is an equilibrium point $\tilde{x}$ of $\tilde{\Gamma}$ with $d(x, \tilde{x}) < \varepsilon$.

Now let $a = (a_\alpha^i)$ be the determining set of Γ. We choose $\eta > 0$ such that

$$4\eta < \min\left\{\frac{1}{m^2}, \frac{\delta_1}{1 + nm^2}\right\},$$

in which $m = m_1 \cdots m_n$. For any game $\bar{\Gamma} \in \mathscr{G}$ with determining set $b = (b_\alpha^i)$, let us suppose that $\mathscr{D}(\Gamma, \bar{\Gamma}) < \eta$, then

$$|N^i(\Gamma) - N^i(\bar{\Gamma})| = \left|\sum_{a,\beta_i,\gamma_i} |a^i_{(a|\beta_i)} - a^i_{(a|\gamma_i)}| - \sum_{a,\beta_i,\gamma_i} |b^i_{(a|\beta_i)} - b^i_{(a|\gamma_i)}|\right|$$

$$\leqslant \sum_{a,\beta_i,\gamma_i} |a^i_{(a|\beta_i)} - b^i_{(a|\beta_i)}| + \sum_{a,\gamma_i,\gamma_i} |a^i_{(a|\gamma_i)} - b^i_{(a|\gamma_i)}|$$

$$= 2m_i \sum_{a,\beta_i} |a^i_{(a|\beta_i)} - b^i_{(a|\beta_i)}|$$

---

1 The concept of essential fixed point with respect to C_0 introduced here is a little more general than the concept of essential fixed point (with respect to $C(S^*)$) introduced in [2] and [3] so far as one-valued mappings are concerned. The Lemma specializes the corresponding result in [2] and [3] when $C_0 = C(S^*)$ (for one-valued mappings). But, it is easy to verify that for any complete subspace $C_0 \subset C(S^*)$, the reasonings in [3] remain valid for our case. Moreover, Lemma 8 can also be extended to multivalued mappings in an evident manner.

$$= 2m_i^2 \sum_a |a_a^i - b_a^i|$$

$$\leqslant 2m^2 \mathscr{D}(\Gamma, \bar{\Gamma}) < 2m^2 \eta < \frac{1}{2}.$$

Since $N^i(\Gamma) = 1$, we get

$$|1 - N^i(\bar{\Gamma})| \leqslant 2m^2 \mathscr{D}(\Gamma, \bar{\Gamma}) < \frac{1}{2} \quad (i \in I).$$

It follows that $N^i(\bar{\Gamma}) > 1/2 (i \in I)$. Hence, the determining set $C = (c_a^i)$ of the normalization $\widetilde{\Gamma}$ of the game $\bar{\Gamma}$ is given by

$$c_a^i = \frac{1}{N^i(\bar{\Gamma})} \left(b_a^i - \frac{1}{m_i} \sum_{\beta_i} b_{(a|\beta_i)}^i \right).$$

Therefore,

$$\mathscr{D}(\Gamma, \widetilde{\Gamma}) = \sum_{i,a} |a_a^i - c_a^i|$$

$$= \sum_{i,a} \left| a_a^i - \frac{b_a^i - \frac{1}{m_i} \sum_{\beta_i} b_{(a|\beta_i)}^i}{N^i(\bar{\Gamma})} \right|$$

$$\leqslant 2 \sum_{i,a} \left| a_a^i N^i(\bar{\Gamma}) - b_a^i + \frac{1}{m_i} \sum_{\beta_i} b_{(a|\beta_i)}^i \right|$$

$$= 2 \sum_{i,a} \left| a_a^i - b_a^i + (N^i(\bar{\Gamma}) - 1)a_a^i \right.$$

$$\left. + \frac{1}{m_i} \sum_{\beta_i} (b_{(a|\beta_i)}^i - a_{(a|\beta_i)}^i) \right|;$$

the last step follows from $P_{(a||i)}^i(\Gamma) = 0$. Furthermore, since

$$1 = N^i(\Gamma) = \sum_{a,\beta_i,\gamma_i} |a_{(a|\beta_i)}^i - a_{(a|\gamma_i)}^i|$$

$$\geqslant \sum_{a,\beta_i} \left| \sum_{\gamma_i} (a_{(a|\beta_i)}^i - a_{(a|\gamma_i)}^i) \right|$$

$$= \sum_{a,\beta_i} m_i |a_{(a|\beta_i)}^i| = m_i^2 \sum_a |a_a^i|,$$

we have
$$\sum_{i,a} |a_a^i| \leqslant n.$$

Hence,
$$\mathscr{D}(\Gamma, \widetilde{\Gamma}) \leqslant 4 \sum_{i,a} |a_a^i - b_a^i| + 2 \sum_{i,a} |1 - N^i(\bar{\Gamma})| \cdot |a_a^i|$$
$$\leqslant 4\mathscr{D}(\Gamma, \bar{\Gamma}) + 4nm^2 \mathscr{D}(\Gamma, \bar{\Gamma})$$
$$< 4(1 + nm^2)\eta < \delta_1.$$

Thus, there is an equilibrium point $\widetilde{x}$ of $\widetilde{\Gamma}$ with $d(x, \widetilde{x}) < \varepsilon$. Since $\bar{\Gamma} \sim \widetilde{\Gamma}$, $\widetilde{x}$ is also an equilibrium point of $\bar{\Gamma}$ by Lemma 2. Therefore, for the equilibrium point x of the game $\Gamma \in \mathscr{G}^*$ and every $\varepsilon > 0$, there is an $\eta > 0$ such that for any $\bar{\Gamma} \in \mathscr{G}$ with $\mathscr{D}(\Gamma, \bar{\Gamma}) < \eta$, there is an equilibrium point $\widetilde{x}$ of $\bar{\Gamma}$ with $d(x, \widetilde{x}) < \varepsilon$. This proves that x is an essential equilibrium point of Γ, and the proof is complete.

Lemma 10 *For every game $\Gamma_a \in \mathscr{G}^*$ and every $\varepsilon > 0$, there exists an essential game $\Gamma_b \in \mathscr{G}^*$ such that*
$$\mathscr{G}(\Gamma_a, \Gamma_b) < \varepsilon.$$

Proof. For any game $\Gamma_b \in \mathscr{G}$ with $\mathscr{G}(\Gamma_a, \Gamma_b) < \varepsilon$, as
$$|N^i(\Gamma_a) - N^i(\Gamma_b)| \leqslant 2m^2 \mathscr{D}(\Gamma_a, \Gamma_b)$$
and all $N^i(\Gamma_a) = 1$, we have $N^i(\Gamma_b) \neq 0$ whenever ε is small enough.

As $\mathscr{G}_0$ and C_0 are homoeomorphic, for $\Gamma_a \in \mathscr{G}^*$ and $\varepsilon > 0$, there is a $\delta > 0$ such that for any $\Gamma_b \in \mathscr{G}_0$ with $\rho(f_{\Gamma_a}, f_{\Gamma_b}) < \delta$, we have $\mathscr{G}(\Gamma_a, \Gamma_b) < \varepsilon$. By Lemma 8, there is $\Gamma_b \in \mathscr{G}_0$ with $\rho(f_{\Gamma_a}, f_{\Gamma_b}) < \delta$ such that $f_{\Gamma_b} \in C_0$ is essential in C_0. It follows that $\mathscr{D}(\Gamma_a, \Gamma_b) < \varepsilon$. If ε is small enough, then $N^i(\Gamma_b) = 1$, i. e., $\Gamma_b \in \mathscr{G}^*$. By Lemma 9, Γ_b is an essential game, and the proof is complete.

Proof of Theorem A. Let $\Gamma_a \in \mathscr{G}$ be an arbitrary game. For every $\varepsilon > 0$, first and foremost, there is a game $\Gamma_b \in \mathscr{G}$ with all $N^i(\Gamma_b) \neq 0$ such that $\mathscr{D}(\Gamma_a, \Gamma_b) < \varepsilon/2$. Thus, the unique normalization $\Gamma_a \in \mathscr{G}_0$ of Γ_b is determined by

$$c_a^i = \frac{1}{N^i(\Gamma_b)} \left(b_a^i - \frac{1}{m_i} P_{(a||i)}^i(\Gamma_b) \right). \tag{14}$$

Since all $N^i(\Gamma_c) = 1$, we have $\Gamma_c \in \mathscr{G}^*$.

Now let the numbers $N^i(\Gamma_b)$ and $P^i_{(a||i)}(\Gamma_b)$ in (14) be fixed and let $b = (b^i_a)$ vary in $\mathscr{G}$, then (14) is a topological transformation T of $\mathscr{G}$ onto itself, and $\Gamma_c = T(\Gamma_b)$. Therefore, for $\varepsilon > 0$, there is a $\delta > 0$ such that for any $\Gamma \in \mathscr{G}$ with $\mathscr{D}(\Gamma_c, \Gamma) < \delta$, we have $\mathscr{D}(\Gamma_b, T^{-1}(\Gamma)) << \varepsilon/2$. If Γ is an essential game, then $T^{-1}(\Gamma)$ is also essential since $\Gamma \sim T^{-1}(\Gamma)$

Since $\Gamma_c \in \mathscr{G}^*$, there is by Lemma 10 an essential game $\Gamma_d \in \mathscr{G}^*$ such that $\mathscr{D}(\Gamma_c, \Gamma_d) < \delta$. Let $\Gamma_e = T^{-1}(\Gamma_d)$. Then Γ_e is an essential game with $\mathscr{D}(\Gamma_b, \Gamma_e) < s/2$. Therefore, for every $\Gamma_a \in \mathscr{G}$ and every $\varepsilon > 0$. there is an essential game $\Gamma_e \in \mathscr{G}$ such that
$$\mathscr{D}(\Gamma_a, \Gamma_e) < \varepsilon,$$
and the proof is complete.

Proof of Theorem B. For any game $\Gamma \in \mathscr{G}$ having only finite number of equilibrium points, $f_r \in C(S^*)$ has only finite number of fixed points by Lemma 1. By a theorem of Fort ([2], Theorem 3), there is at least one essential fixed point $x \in S^*$ of f_r with respect to $C(S^*)$: for every $\varepsilon > 0$, there is a $\delta > 0$ $d(x, y) < \varepsilon$ such that for any $g \in C(S^*)$ with $\rho(f_r, g) < \delta$, there is a fixed point y of g with $d(x, y) < \varepsilon$. By Lemma 4, there is a $\delta_1 > 0$ such that for any $\widetilde{\Gamma} \in \mathscr{G}$ with $\mathscr{D}(\Gamma, \widetilde{\Gamma}) < \delta_1$, we have $\rho(f_r, f_{\widetilde{r}}) < \delta$. Hence, there is a fixed point y of $f_{\widetilde{r}}$, i.e., an equilibrium point y of $\widetilde{\Gamma}$ such that $d(x, y) < \varepsilon$. Thus, x is an essential equilibrium point of Γ, and the proof is complete.

References

[1] Nash, J. Non-cooperative games. *Annals of Math.*, 1951, 54: 286-295.

[2] Fort, Jr., M. K. Essential and non-essential fixed points. *American Journal of Math.*, 1950, 72: 315-322.

[3] Jiang, Jia-he. Essential fixed point of the multivalued mappings. *Scientia Sinica*, 1962, 11: 293-298.

具有对偶有理分割的代数簇 *

以下我们大体上是用 A. Weil[8] 一书中的词汇, 因而所谓"簇"或"代数簇"是指 Weil 意义下的"抽象簇"(abstract variety). 我们并将使用 Zariski 拓扑, 文中一切拓扑名称, 如开集、闭集, 都指对此拓扑而言. 凡此都不再解释.

§1. 设 V 是一个没有奇点的代数簇, 命 $G(V,s)$ 是 V 上一切 s 维循环 (cycle) 所成的群. 又命 $G_n(V,s), G_r(V,s), G_a(V,s)$ 依次为 $G(V,s)$ 中数值等价于 0, 有理等价于 0 与代数等价于 0 的循环所成的子集合. 我们知道, 这些集合都是 $G(V,s)$ 的子群, 而有

$$G(V,s) \supset G_n(V,s) \supset G_a(V,s) \supset G_r(V,s).$$

因而有自然同态

$$\alpha_* : G(V,s)/G_r(V,s) \to G(V,s)/G_a(V,s),$$

$$\beta_* : G(V,s)/G_a(V,s) \to G(V,s)/G_n(V,s),$$

其中诸商群各称为 V 的 s 维有理等价群、代数等价群与数值等价群, 诸群中的傍系称为有理等价系等. 同一有理等价系中的循环称为互相有理等价等 [1]. 如果 $G(V,s)$ 中有有限多个循环 $X_1, \cdots, X_s$ 使 $G(V,s)$ 中任一循环都有理 (或代数) 等价于 $X_1, \cdots, X_s$ 的一个 (整系数) 线性组合, 则称 $X_1, \cdots, X_s$ 是 $G(V,s)$ 的一组有限有理基 (或代数基). 如果 $X_1, \cdots, X_s$ 的任一线性组合都不有理 (或代数) 等价于 0, 则称它们为一组最小基. 在以上诸商群中, 对代数几何来说最重要的是代数等价群 $G(V,s)/G_a(V,s)$. 它的确定以及是否存在一组最小有限基的问题是重要而极困难的, 没有一般的解决方法 (见 Hodge [4] XII §11). 但在周炜良 [2] 中, 证明了对于某类代数簇可以给出一组有限有理基. [2] 事实上也提供了一个简单而一般的方法, 对于像 Ehresmann 在 [3] 中所论述的那一类代数簇以及 Hodge 在 [4, XII §2, 11 与 XII, XIV] 中所讨论的那种代数簇可以完全确定它们的代数等价群与有理等价群. 本文就是详述此点.

---

* 本文原载《数学进展》, 1965 年, 第 8 卷第 4 期, 402–409.

1) 以上这些概念的一般定义可参阅 Weil[9] 的 § 10 (关于代数等价, V 无任何限制) 与 Samuel [6](关于有理等价, 投影簇的限制是不必要的). 另一种定义方式可参阅 Chow(周炜良)[2](关于有理等价, V 无任何限制) 与 Hodge[4] XII. 关于代数等价, 特征 0 的限制是不必要的. 对于没有奇点的投影簇来说, 这两种定义方式是一致的, 它们的证明可参阅 Samuel [6](关于有理等价) 与 Igusa[5](关于代数等价, 复域的限制也是不必要的). 这方面的一般介绍可参阅 Baldassari[1] VI 与 Samuel [7].

依周炜良 [2], 设 V 是代数簇, $\{W_i\}$ 是包含于 V 的有限多个代数簇, 使 V 的每一点在一个且恰在一个 W_i 中. 于是 $\{W_i\}$ 称为 V 的一个分割 (dissection), W_i 称为这个分割的胞腔. 如果每一个 W_i 都可在一处处双正则的双有理变换下等价于一仿射空间, 则这个分割称为有理的 (rational). 于是有下述

周炜良定理 如果代数簇 V 有有理分割, 则这个分割的所有 s 维胞腔的闭包作成 $G(V,s)$ 中的一个有限有理基.

今设 V 是没有奇点的投影簇, 因而 V 上可建立交载理论 (有理等价情形见 Samuel [6], 代数等价情形见 Hodge [4, XII], 其中特征 0 的限制是不必要的). 我们将引入下述概念:

定义 假设 V 上有两个有理分割 $\{W_i\}$ 与 $\{W_i^*\}$, 在它们的胞腔之间有一一对应 $W_i \leftrightarrow W_i^*$, 使 (i) $\dim W_i + \dim W_i^* = \dim V$; (ii) 如果 $\dim W_i + \dim W_j^* < n$ 或 $\dim W_i + \dim W_j^* = n$ 但 $i \neq j$, 则 W_i 与 W_j^* 不相遇; (iii) W_i 与 W_i^* 单一相交于一点[1]), 于是 $\{W_i\}$ 与 $\{W_j^*\}$ 将称为 V 的一组对偶有理分割.

作为周炜良定理的一个简单补充, 我们有

定理 如果没有奇点的投影簇 V 有一组对偶的有理分割 $\{W_i\}$ 与 $\{W_j^*\}$, 则

$$\alpha_*: G(V,s)/G_r(V,s) \approx G(V,s)/G_a(V,s),$$

$$\beta_*: G(V,s)/G_a(V,s) \approx G(V,s)/G_n(V,s),$$

且分割 $\{W_i\}$ (或 $\{W_j^*\}$) 中所有 s 维胞腔的闭包作成 $G(V,s)$ 中的一组最小代数基, 同时也是最小有理基. 如果 V 以复数域为一定义域, 因而可视 V 为一复流形时, 下述自然同态也是一个同构:

$$\gamma_*: G(V,s)/G_a(V,s) \approx H_{2s}(V),$$

这里 H_{2s} 指 $2s$ 维下同调群 (整系数).

证明是很容易的, 设 $W_{i_1}, \cdots, W_{i_k}$ 是 $\{W_i\}$ 中 s 维胞腔的全体, 而 $X \in G_n(V,s)$, 则依周炜良定理, 应有整数 $n_{i_1}, \cdots, n_{i_k}$, 使 X 有理等价于循环 $W = n_{i_1}\overline{W}_{i_1} + \cdots + n_{i_k}\overline{W}_{i_k}$.

以 $[X]$ 表循环 X 的有理等价系, 则应有

$$0 = [X] \cdot [W_{i_j}^*] = [W] \cdot [W_{i_j}^*] = n_{i_j},$$

故有 $W = 0$ 而 X 有理等价于 0, 即 $G_n(V,s) \subset G_r(V,s)$. 因之有 $G_n(V,s) = G_a(V,s) = G_r(V,s)$, 而定理的前二部分得证. 末一部分也是显然的.

---

1) 即在该点的相交指数有定义且等于 1. 下同.

注. 从证明可以看出, 要定理的结论成立, 对偶有理分割存在的要求是过强的. 事实上, 只要求有有理分割 $\{W_i\}$ 且对每一 W_i 来说都存在一循环 $\dim W_i + \dim Z_i = \dim V$ 使 Z_i 与 $\overline{W}_i (= W_i$ 的闭胞) 的相交指数为 1, 而 Z_i 与任一 $\overline{W}_j (= W_j$ 的闭胞, $\dim W_j = \dim W_i, W_j \neq W_i)$ 的相交指数为 0 就可以了. 但由于符合这些条件而又比较熟悉的代数簇像以下将要论证的那样, 都具有对偶有理分割, 故归纳成上述定理.

§2. 命 $P_n = P_n(\tilde{K})$ 为万有域 (universal domain) $\tilde{K}$ 上的 n 维投影空间, 而以 $[r], [r]', [r]_1, \{r\}$ 等表 P_n 中的 r 维投影子空间. 对整数 $0 \leqslant \alpha < \beta \leqslant n$, 命 $\tilde{\Omega}^n_{\alpha,\beta}$ 为所有偶 $([\alpha], [\beta])$ 组成的代数簇, 这里 $[\alpha] \subset [\beta] \subset [n]$. 由于例如 $\alpha = 0$ 时的 $\tilde{\Omega}^n_{0,\beta}$ 对于代数簇上 (陈省身) 示性系研究的重要性 (见 [10]), 我们将应用 §1 所述方法比较详细地叙述 $\tilde{\Omega}^n_{\alpha,\beta}$ 的代数等价群的确定. Ehresmann[3] 提到的其他代数簇, 也可用同样办法确定出来. 但由于缺少应用, 本文中将略而不论.

在 P_n 中取一固定的 $\{n-\alpha-1\}$, 命 Ω^n_α 与 $\Omega^{n-\alpha-1}_{\beta-\alpha-1}$ 各为 P_n 与 $\{n-\alpha-1\}$ 中所有 α 维与 $\beta-\alpha-1$ 维投影子空间所成的 Grassmann 簇, 又命 k_0 为 $\{n-\alpha-1\}$ 与 $\Omega^n_\alpha, \Omega^{n-\alpha-1}_{\beta-\alpha-1}, \tilde{\Omega}^n_{\alpha,\beta}$ 的一个公共定义域, 于是可定义 $U = \tilde{\Omega}^n_{\alpha,\beta}$ 与 $V = \Omega^n_\alpha \times \Omega^{n-\alpha-1}_{\beta-\alpha-1}$ 间的一个代数对应 T 如下: 命 $(\{\alpha\}, \{\beta\})$ 为 $U = \tilde{\Omega}^n_{\alpha,\beta}$ 对 k_0 的一个一般点, 则 $\{\beta\}$ 与 $\{n-\alpha-1\}$ 的交是一 $\{\beta-\alpha-1\}$, 而为 $\Omega^{n-\alpha-1}_{\beta-\alpha-1}$ 对 k_0 的一个一般点. 对应 T 即定义为 $U \times V$ 中以 $P = (\{\alpha\}, \{\beta\}), (\{\alpha\}, \{\beta-\alpha-1\})$ 为对 k_0 的一般点的子簇.

引理 1 T 是 U 与 V 间的一个双有理对应.

证. 设 $[\alpha], [\beta-\alpha-1]$ 各为 $\Omega^n_\alpha, \Omega^{n-\alpha-1}_{\beta-\alpha-1}$ 对 k_0 的一般点, 并设 $P' = (([\alpha]', [\beta]'), ([\alpha], [\beta-\alpha-1])) \in T$, 则 P' 是 P 对于 k_0 的一个特定化. 于是 $([\alpha]', [\beta]', [\alpha], [\beta-\alpha-1])$ 是 $(\{\alpha\}, \{\beta\}, \{\alpha\}, \{\beta-\alpha-1\})$ 对 k_0 的一个特定化, 而有 $[\alpha]' = [\alpha]$. 又 $[\alpha]$ 作为 P_n 中对 k_0 的一般 α 维投影子空间不能与 $\{n-\alpha-1\}$ 相交, 因而 $[\alpha]$ 与 $[\beta-\alpha-1]$ 的联合是一 $[\beta]$. 由于 $\{\beta\}$ 是 $\{\alpha\}$ 与 $\{\beta-\alpha-1\}$ 的联合, 故在特定化下 $[\beta]'$ 也是 $[\alpha]$ 与 $[\beta-\alpha-1]$ 的联合, 即 $[\beta]' = [\beta]$. 因之, 当将 $U \times V$ 投影于 V 上时, T 中投影于 $([\alpha], [\beta-\alpha-1]) \in V$ 的点是唯一确定的, 即 $[T:V] = 1$ 或 T 正则地投影于 V 之上. 同样可证 $[T:U] = 1$ 或 T 正则地投影于 U 之上. 因而 T 是 U 与 V 间的一个双有理对应.

今依 Ehresmann[3] 中的方法并采用它的符号, 在 P_n 中取一固定序列

(A) $P_n = [n] \supset [n-1] \supset \cdots \supset [1] \supset [0]$.

对序列 (A) 而言, 命

$$F = \begin{bmatrix} a_{i_0} \cdots a_{i_\alpha} \\ a_0 \cdots a_\beta \end{bmatrix} \quad (0 \leqslant a_0 < a_1 < \cdots < a_\beta \leqslant n, 0 \leqslant i_0 < \cdots < i_\alpha \leqslant \beta) \tag{1}$$

表所有 $([\alpha], [\beta]) \in \tilde{\Omega}^n_{\alpha,\beta}$ 所成的点集, 满足条件

$$\left.\begin{array}{l}\dim([\beta] \cap [a_i]) \geqslant i, \\ \dim([\alpha] \cap [a_{i_j}]) \geqslant j.\end{array}\right\} \quad (2)$$

一般说来, 命

$$\begin{bmatrix} a'_0 \cdots a'_\alpha \\ a_0 \cdots a_\beta \end{bmatrix}$$

表所有 $([\alpha], [\beta]) \in \tilde{\Omega}^n_{\alpha, \beta}$ 所成的点集, 满足条件

$$\left.\begin{array}{l}\dim([\beta] \cap [a_i]) \geqslant i, \\ \dim([\alpha] \cap [a'_j]) \geqslant j.\end{array}\right\}$$

于是依 Ehresmann [3, p. 436] 上所证, 在 $a' \neq$ 任一 a_i 时, 这个点集是形如 (1) 的点集的和集. 对于 (1) 那样的 F 而言, 则有

引理 2 F 是 Weil 意义下的代数簇, 其维数为

$$\rho = \sum_{k=0}^{\beta}(a_k - k) + \sum_{j=0}^{\alpha}(i_j - j). \quad (3)$$

证. 命 K 是 (A) 中诸簇的一个公共定义域, 在 $[a_0]$ 中取一对 K 的一般点 $\eta^{(0)}$, 又在 $[a_1]$ 中取一对 $K(\eta^{(0)})$ 的一般点 $\eta^{(1)}$, 依次类推, 直至在 $[a_\beta]$ 中取一对 $K(\eta^{(0)}, \cdots, \eta^{(\beta-1)})$ 的一般点 $\eta^{(\beta)}$, 则诸点 $\eta^{(i)}$ 必线性无关. 因而 $\eta^{(0)}, \cdots, \eta^{(i)}$ 确定了一个 i 维投影空间 B_i, 特别命 $B_\beta = B$, 依次在 $B \cap [a_{i_j}] = B_{i_j}$ 中取一对 $K(\eta^{(0)}, \cdots, \eta^{(i_j)}, \xi^{(0)}, \cdots, \xi^{(j-1)})$ 的一般点 $\xi^{(j)}(j = 0, 1, \cdots, \alpha)$, 则诸点 $\xi^{(j)}$ 也必线性无关. 因而 $\xi^{(0)}, \cdots, \xi^{(j)}$ 确定一个 j 维投影空间 A_j, 特别命 $A_\alpha = A$. 我们将证 (A, B) 是 F 的一个一般点, 由此知 F 是代数簇.

为此, 先设 $(\overline{A}, \overline{B})$ 是 (A, B) 对 K 的一个特定化, 于是从 $A \subset B$, $\dim(B \cap [a_i]) = \dim B_i = i$ 以及 $\dim(A \cap [a_{i_j}]) = \dim A_j = !j$ 可得 $\overline{A} \subset \overline{B}$, $\dim(\overline{B} \cap [a_i]) \geqslant i$ 以及 $\dim(\overline{A} \cap [a_{i_j}]) \geqslant j$. 因而 $(\overline{A}, \overline{B}) \in F$. 反之, 设 $(\overline{A}, \overline{B}) \in F$, 则有 $\overline{A} \subset \overline{B}$ 与 $\dim(\overline{B} \cap [a_i]) \geqslant i, \dim(\overline{A} \cap [a_{i_j}]) \geqslant j$. 因而可在 $\overline{B} \cap [a_i]$ 中取点 $\overline{\eta}^{(i)}$, 使 $\overline{\eta}^{(i)}$ 线性独立于 $\overline{\eta}^{(0)}, \cdots, \overline{\eta}^{(i-1)}(i = 1, \cdots, \beta)$; 又在 $\overline{A} \cap [a_{i_j}]$ 中取点 $\overline{\xi}^{(j)}$, 使 $\overline{\xi}^{(j)}$ 线性独立于 $\overline{\xi}^{(0)}, \cdots, \overline{\xi}^{(j-1)}(j = 1, \cdots, \alpha)$, 于是 $\overline{B}$ 为 $\overline{\eta}^{(0)}, \cdots, \overline{\eta}^{(\beta)}$ 所张的 β 维投影空间, 而 $\overline{A}$ 为 $\overline{\xi}^{(0)}, \cdots, \overline{\xi}^{(\alpha)}$ 所张的投影空间. 由前面 $\eta^{(i)}$ 与 $\xi^{(j)}$ 的选择, $\overline{\eta}^{(i)}$ 是 $\eta^{(i)}$ 对 K 的一个特定化, 而 $\overline{\xi}^{(j)}$ 是 $\xi^{(j)}$ 对 K 的一个特定化. 由于诸 $\eta^{(i)}$ 与 $\xi^{(j)}$ 是对 K 独立的, 且每个 $K(\eta^{(i)})$ 与 $K(\xi^{(j)})$ 都是 K 上的纯超越扩充, 因而也是正则扩充, 故 $(\overline{\xi}^{(0)}, \cdots, \overline{\xi}^{(\alpha)}, \overline{\eta}^{(0)} \cdots \overline{\eta}^{(\beta)})$ 是 $(\xi^{(0)}, \cdots, \xi^{(\alpha)}, \eta^{(0)}, \cdots, \eta^{(\beta)})$ 对 K 的一个特定化 (A. Weil [8, II §1 定理 5]. 由于 $\xi^{(0)}, \cdots, \xi^{(\alpha)}$ 与 $\eta^{(0)}, \cdots, \eta^{(\beta)}$ 各张成 A 与 B, 故有特定化 $(A, B) \to (\overline{A}, \overline{B})$. 因而 (A, B) 是 F 对 K 的一般点, 如所欲证.

其次证 (3) 式. 试考虑 $P_{\alpha_{i_0}} \times \cdots \times P_{\alpha_{i_\alpha}} \times P_{\alpha_0} \times \cdots \times P_{\alpha_\beta}$ 与 $\tilde{\Omega}_{\alpha,\beta}^n$ 间由 $(\xi^{(0)}, \cdots, \xi^{(\alpha)}, \eta^{(0)}, \cdots, \eta^{(\beta)}; A, B)$ 所定的不可约代数对应 C. 应用计数定理, 即得

$$\sum_{j=0}^{\alpha} i_j + \sum_{i=0}^{\beta} a_i = \rho + \sum_{i=0}^{\beta} i + \sum_{j=0}^{\alpha} j,$$

也就是 (3) 式. 引理 2 证毕.

今命 F 同 (1), 而命 F^- 为从 F 中除去所有有意义的 $\begin{bmatrix} b'_0 \cdots b'_\alpha \\ b_0 \cdots b_\beta \end{bmatrix}$ 而成的代数簇, 这里 $0 \leqslant b_0 \leqslant b_1 < \cdots < b_\beta \leqslant n, 0 \leqslant b'_0 < \cdots < b'_\alpha \leqslant n, b_i \leqslant a_i (i = 0, \cdots, \beta), b'_j \leqslant a_{i_j} (j = 0, \cdots, \alpha)$, 至少有一 $b'_j < a_{i_j}$ 或 $b_i < a_i$.

引理 3 F^- 可在一处处双正则双有理变换下等价于一仿射空间.

证. 由于对不同序列 (A) 所定义的代数簇 F 与 F^- 彼此投影等价, 因之只要适当选择序列 (A) 来证明引理 3 对相应的 F 成立即足.

为此取一 P_n 中固定的 $\{n - \alpha - 1\}$, 并定义 $\Omega_\alpha^n, \Omega_{\beta-\alpha-1}^{n-\alpha-1}$ 与 T 如前, 而序列 (A) 如下: 首先, 在 $\{n - \alpha - 1\}$ 中任选一序列

(B) $\{n - \alpha - 1\} = [n - \alpha - 1]' \supset [n - \alpha - 2]' \supset \cdots \supset [1]' \supset [0]'$.

其次在 P_n 中依次取 $[a_0] \subset [a_1] \subset \cdots \subset [a_\beta]$ 使 ($\circ$ 表联合)

$$[a_0] = [a_0]', \cdots, [a_{i_0} - 1] = [a_{i_0} - 1]', \tag{4}$$

$$[a_{i_0}] \cap \{n - \alpha - 1\} = [a_{i_0} - 1]', \tag{5}_0$$

$$[a_{i_0+1}] = [a_{i_0}] \circ [a_{i_0+1} - 1]', \cdots, [a_{i_1-1}] = [a_{i_0}] \circ [a_{i_1-1} - 1]', \tag{6}_0$$

$$[a_{i_1}] \cap \{n - \alpha - 1\} = [a_{i_1} - 2]', \tag{5}_1$$

$$[a_{i_1+1}] = [a_{i_1}] \circ [a_{i_1+1} - 2]', \cdots, [a_{i_2-1}] = [a_{i_1}] \circ [a_{i_2-1} - 2]', \tag{6}_1$$

$$[a_{i_2}] \cap \{n - \alpha - 1\} = [a_{i_2} - 3]', \tag{5}_2$$

$$\cdots\cdots\cdots\cdots\cdots\cdots\cdots\cdots\cdots$$

$$[a_{i_\alpha}] \cap \{n - \alpha - 1\} = [a_{i_\alpha} - \alpha - 1]', \tag{5}_a$$

$$[a_{i_\alpha+1}] = [a_{i_\alpha}] \circ [a_{i_\alpha+1} - \alpha - 1]', \cdots, [a_\beta] = [a_{i_\alpha}] \circ [a_\beta - \alpha - 1]'. \tag{6}_a$$

最后将上述序列 $[a_0] \subset \cdots \subset [a_\beta]$ 扩充为序列 (A), 使其中

$$[a_{i_1} - 1] = [a_{i_0}] \circ [a_{i_1} - 2]', [a_{i_2} - 1] = [a_{i_1}] \circ [a_{i_2} - 3]', \cdots, [a_{i_a} - 1]$$
$$= [a_{i_{a-1}}] \circ [a_{i_a} - \alpha - 1]', \tag{7}$$

$$[a_{i_0+1} - 1] = [a_{i_0}] \circ [a_{i_0+1} - 2]', \cdots, [a_{i_1-1} - 1] = [a_{i_0}] \circ [a_{i_1-1} - 2]', \tag{8}_0$$

$$[a_{i_1+1} - 1] = [a_{i_1}] \circ [a_{i_1+1} - 3]', \cdots, [a_{i_2-1} - 1] = [a_{i_1}] \circ [a_{i_2-1} - 3]', \tag{8}_1$$

$$[a_{i_\alpha+1}-1]=[a_{i_\alpha}]\circ[a_{i_\alpha+1}-\alpha-2]',\cdots,[a_\beta-1]=[a_{i_\alpha}]\circ[a_\beta-\alpha-2]'. \qquad (8)_\alpha$$

此外任意.

今对 (A) 定义 F 与 F^-, 又对 (A) 定义 Ω_α^n 中的 Schubert 簇

$$F_\alpha=[a_{i_0}a_{i_1}\cdots a_{i_\alpha}].$$

对 (B) 定义 $\Omega_{\beta-\alpha-1}^{n-\alpha-1}$ 中的 Schubert 簇

$$F_\beta=[a_0\cdots a_{i_0-1},a_{i_0+1}-1,\cdots,a_{i_1-1}-1,a_{i_1+1}-2,\cdots,$$
$$a_{i_2-1}-2,\cdots,a_{i_\alpha+1}-\alpha-1,\cdots,a_\beta-\alpha-1].$$

命 F_α^- 为从 F_α 中除去所有有意义的子簇 $[b_{i_0}b_{i_1}\cdots b_{i_\alpha}]$ 后所余的簇, 这里 $b_{i_j}\leqslant a_{i_j}, j=0,\cdots,\alpha$, 而至少有一 $b_{i_j}<a_{i_j}$. 同样命 F_β^- 为从 F_β 中除去所有有意义的子簇 $[b_0\cdots b_{i_0-1},b_{i_0+1}-1,\cdots,b_{i_1-1}-1,\cdots,b_{i_\alpha+1}-\alpha-1,\cdots,b_\beta-\alpha-1]$ 后所余的簇, 这里 $b_0\leqslant a_0,\cdots,b_{i_0-1}\leqslant a_{i_0-1},\cdots,b_{i_\alpha+1}\leqslant a_{i_\alpha+1},\cdots,b_\beta\leqslant a_\beta$, 而其中至少有一个是 $<$ 号. 由 Grassmann 簇的已知结果 (Hodge-Pedoe, [4]XIV), F_α^- 与 F_β^- 都双正则双有理等价于同维数的仿射空间. 因之要证明引理 3, 只须证明在双有理变换 T 下, F^- 处处双正则地变为 $F_\alpha^-\times F_\beta^-$ 即可.

为此设 $(\{\alpha\},\{\beta\})\in F^-$. 由 (4) 与 $(5)_0$ 可见, $\dim(\{\alpha\}\cap[a_{i_0}])=0$ 且 $\{\alpha\}\cap[a_{i_0}]\cap\{n-\alpha-1\}=\varnothing$, 否则 $(\{\alpha\},\{\beta\})$ 将属于 $\begin{bmatrix}a_{i_0}-1, & a_{i_1}\cdots a_{i_\alpha}\\ a_0\cdots a_\beta\end{bmatrix}$, 而不属于 F^-. 其次, 由 $(5)_1$ 与 (7) 可见, $\dim(\{\alpha\}\cap[a_{i_1}])=1$ 且 $\{\alpha\}\cap[a_{i_1}]\cap\{n-\alpha-1\}=\varnothing$, 否则 $(\{\alpha\},\{\beta\})$ 将 $\in\begin{bmatrix}a_{i_0},a_{i_1}-1, & a_{i_2}\cdots a_{i_\alpha}\\ a_0\cdots a_\beta\end{bmatrix}$ 而 $\in F^-$. 依次类推, 可知 $\dim(\{\alpha\}\cap[a_{i_j}])=j(j=0,1,\cdots,\alpha)$ 而 $\{\alpha\}\cap\{n-\alpha-1\}=\{\alpha\}\cap[a_{i_\alpha}]\cap\{n-\alpha-1\}=\varnothing$. 于是 $\{\beta\}\cap\{n-\alpha-1\}$ 恰为一 $\{\beta-\alpha-1\}$. 再由 (6) 可知, 此 $\{\beta-\alpha-1\}$ 必 $\in F_\beta$. 最后由 (8) 知, 此 $\{\beta-\alpha-1\}\in F_\beta^-$. 综述之, T 中有唯一一点 $((\{\alpha\},\{\beta\}),(\{\alpha\},\{\beta-\alpha-1\}))$, 它在第一因子上的投影为 $(\{\alpha\},\{\beta\})$, 在第二因子上的投影为 $(\{\alpha\},\{\beta-\alpha-1\})\in F_\alpha^-\times F_\beta^-$, 而 $(\{\alpha\},\{\beta-\alpha-1\})$ 是 $(\{\alpha\},\{\beta-\alpha-1\})$ 在 T 下的唯一对应点. 而且由于 $\{\beta-\alpha-1\}=\{\beta\}\cap\{n-\alpha-1\}$, 对应 T 在 $(\{\alpha\},\{\beta\})$ 处是正则的.

同样可知, 对任一 $(\{\alpha\},\{\beta-\alpha-1\})\in F_\alpha^-\times F_\beta^-$, 必有 $\{\alpha\}\cap\{n-\alpha-1\}=\varnothing$, 且置 $\{\alpha\}\circ\{\beta-\alpha-1\}=\{\beta\}$ 时, $(\{\alpha\},\{\beta\})$ 是在 T^{-1} 下唯一对应于 $(\{\alpha\},\{\beta-\alpha-1\})$ 的点, 并在该处正则. 由此知, T 处处双正则双有理地变 F^- 为 $F_\alpha^-\times F_\beta^-$. 如所欲证.

定理 $\tilde\Omega_{\alpha,\beta}^n$ 具有对偶的有理分割.

证. 对一固定的序列 (A), 定义基本簇

$$F_\lambda = \begin{bmatrix} a_{i_0} & \cdots & a_{i_\alpha} \\ a_0 & \cdots & a_\beta \end{bmatrix}$$

如前, 这里 λ 跑过某一指数集 Λ. 依引理 2 及其前面附言可见, $\{F_\lambda^-\}_{\lambda \in \Lambda}$ 的全体 $\mathcal{F}$ 构成 $\tilde{\Omega}_{\alpha,\beta}^n$ 的一个分割, 且依引理 3, 这是一个有理分割. 其次再在 P_n 中取一序列

(A)* $P_n = [n]^* \supset [n-1]^* \supset \cdots \supset [0]^*$,

使 $\dim([n-i]^* \cap [i]) = 0 (i = 0, 1, \cdots, n)$. 对 (A)* 定义的基本簇用与以前同样的记号, 而在右上角加 * 号以区别之. 于是对 (A)* 同样有一有理分割 $\{F_\lambda^{*-}\} = \mathcal{F}^*$, 其中与 $\lambda \in \Lambda$ 相对应的基本簇将取作

$$F_\lambda^* = \begin{bmatrix} n - a_{i_\alpha} & \cdots & n - a_{i_0} \\ n - a_\beta & \cdots & n - a_0 \end{bmatrix}^*.$$

我们证明: 在对应 $F_\lambda^- \leftrightarrow F_\lambda^{*-}$ 之下, $\mathcal{F}$ 与 $\mathcal{F}^*$ 构成 $\tilde{\Omega}_{\alpha,\beta}^n$ 的一组对偶有理分割. 对于对偶分割, 定义中的条件 (i) 是自然满足的. 条件 (ii) 依 Ehresmann [3, p.438] 仍然成立. 余下来需要证明条件 (iii), 即对任意 $\lambda \in \Lambda, F_\lambda$ 与 F_λ^* 只相交于一点且相交指数

$$F_\lambda \cdot F_\lambda^* = \begin{bmatrix} a_{i_0} & \cdots & a_{i_\alpha} \\ a_0 & \cdots & a_\beta \end{bmatrix} \cdot \begin{bmatrix} n - a_{i_\alpha} & \cdots & n - a_{i_0} \\ n - a_\beta & \cdots & n - a_0 \end{bmatrix}^* = 1. \tag{9}$$

前者已在上述 Ehresmann [3, p.438] 中证明. 为证 (9') 式, 先注意: 如果易 F_λ, F_λ^* 为任二与之代数等价的代数簇, 只须交截有定义, 它们的相交指数并不因之而变. 因此, 只须选取适当的序列 (A) 与 (A)*, 证明对此定义的相应的 F_λ 与 F_λ^* 的相交指数 $= 1$ 就够了. 为此, 在 P_n 中取两个固定不相交的 $\{n-\alpha-1\}$ 与 $\{\alpha\}$, 并对前者定义双有理变换 T 如引理 1. 今在 $\{n-\alpha-1\}$ 中取序列

(B)$\{n - \alpha - 1\} = [n - \alpha - 1]' \supset \cdots \supset [1]' \supset [0]'$

如前, 又取序列

(B)* $\{n - \alpha - 1\} = [n - \alpha - 1]'^* \supset \cdots \supset [1]'^* \supset [0]'^*$

使当 $j + k < n - \alpha - 1$ 时, $[j]' \cap [k]'^* = \emptyset$. 其次在 $\{\alpha\}$ 中取两序列

(C) $\{\alpha\} = [\alpha]'' \supset [\alpha - 1]'' \supset \cdots \supset [1]'' \supset [0]''$

与

(C)* $\{\alpha\} = [\alpha]''^* \supset [\alpha - 1]''^* \supset \cdots \supset [1]''^* \supset [0]''^*$

使当 $j + k < \alpha$ 时, $[j]'' \cap [k]''^* = \emptyset$. 今取一特殊的序列 (A), 其中诸 $[i]$ 定义如下 (○ 表联合):

$$[0] = [0]', \cdots, [a_{i_0} - 1] = [a_{i_0} - 1]', \tag{10$_0$}$$

$$[a_{i_0}] = [0]'' \circ [a_{i_0} - 1]', \tag{11}_0$$

$$[a_{i_0} + 1] = [0]'' \circ [a_{i_0}]', \cdots, [a_{i_1} - 1] = [0]'' \circ [a_{i_1} - 2]', \tag{10}_1$$

$$[a_{i_1}] = [1]'' \circ [a_{i_1} - 2]', \tag{11}_1$$

$$[a_{i_1} + 1] = [1]'' \circ [a_{i_1} - 1]', \cdots, [a_{i_2} - 1] = [1]'' \circ [a_{i_2} - 3]', \tag{10}_2$$

$$\cdots\cdots\cdots\cdots\cdots$$

$$[a_{i_\alpha}] = [\alpha]'' \circ [a_{i_\alpha} - \alpha - 1]', \tag{11}_\alpha$$

$$[a_{i_\alpha} + 1] = [\alpha]'' \circ [a_{i_\alpha} - \alpha]', \cdots, [n] = [\alpha]'' \circ [n - \alpha - 1]'. \tag{10}_{\alpha+1}$$

同样取一特殊的序列 $(A)^*$, 其中诸 $[i]^*$ 为

$$[0]^* = [0]'^*, \cdots, [n - a_{i_\alpha} - 1]^* = [n - a_{i_\alpha} - 1]'^*, \tag{10}_0^*$$

$$[n - a_{i_\alpha}]^* = [0]''^* \circ [n - a_{i_\alpha} - 1]'^*, \tag{11}_0^*$$

$$[n - a_{i_\alpha} + 1]^* = [0]''^* \circ [n - a_{i_\alpha}]'^*, \cdots, [n - a_{i_{\alpha-1}} - 1]^* \tag{10}_1^*$$
$$= [0]''^* \circ [n - a_{i_{\alpha-1}} - 2]'^*,$$

$$[n - a_{i_{\alpha-1}}]^* = [1]''^* \circ [n - a_{i_{\alpha-1}} - 2]'^*, \tag{11}_1^*$$

$$[n - a_{i_{\alpha-1}} + 1]^* = [1]''^* \circ [n - a_{i_{\alpha-1}} - 1]'^*, \cdots, [n - \alpha_{i_{\alpha-2}} - 1]^*$$
$$= [1]'' \circ [n - a_{i_{\alpha-2}} - 3]'^*, \tag{10}_2^*$$

$$\cdots\cdots\cdots\cdots\cdots$$

$$[n - a_{i_0}]^* = [\alpha]''^* \circ [n - a_{i_0} - \alpha - 1]'^*, \tag{11}_\alpha^*$$

$$[n - a_{i_0} + 1]^* = [\alpha]''^* \circ [n - a_{i_0} - \alpha]'^*, \cdots, [n]^*$$
$$= [\alpha]''^* \circ [n - \alpha - 1]'^*. \tag{10}_{\alpha+1}^*$$

我们将对 (A) 与 $(A)^*$ 定义 F_λ, F_λ^*. 由于条件 (10), (11) 蕴含 (4)—(8), 故依引理 3 的证明, T 在 F_λ^- 上处处双正则地双有理变换 F_λ^- 为 $F_{\lambda\alpha}^- \times F_{\lambda\beta}^-$, 这里

$$F_{\lambda\alpha} = [a_{i_0} a_{i_1} \cdots a_{i_\alpha}],$$
$$F_{\lambda\beta} = [a_0 \cdots a_{i_0-1}, a_{i_0+1} - 1, \cdots, a_{i_1-1} - 1, \cdots, a_{i_\alpha+1} - \alpha - 1, \cdots, a_\beta - \alpha - 1]$$

各为 Ω_α^n 与 $\Omega_{\beta-\alpha-1}^{n-\alpha-1}$ 中对 (A) 与 (B) 定义的 Schubert 簇. 同样 T 也在 F_λ^{*-} 上处处双正则地双有理变换 F_λ^{*-} 为 $F_{\lambda\alpha}^{*-} \times F_{\lambda\beta}^{*-}$, 这里

$$F_{\lambda\alpha}^* = [n - a_{i_\alpha}, \cdots, n - a_{i_0}]^*,$$
$$F_{\lambda\beta}^* = [n - a_\beta, \cdots, n - a_{i_\alpha+1}, \cdots, n - a_{i_1-1} - \alpha, \cdots, n - a_{i_0+1} - \alpha, \cdots,$$
$$n - a_{i_0-1} - \alpha - 1, \cdots, n - a_0 - \alpha - 1]$$

各为 Ω_α^n 与 $\Omega_{\beta-\alpha-1}^{n-\alpha-1}$ 中对 (A)* 与 (B)* 定义的 Schubert 簇. 由 (B) 与 (B)* 的选择, $F_{\lambda\beta}$ 与 $F_{\lambda\beta}^*$ 只有一个交点, 设为 $\{\beta-\alpha-1\}_0$, 且由 Grassmann 簇的已知结果 (Hodge-Pedoe[4], XII), $F_{\lambda\beta}$ 与 $F_{\lambda\beta}^*$ 在 $\{\beta-\alpha-1\}_0$ 处单一相交. 同样, 由 (10), (11), (10)* 与 (11)* 可知, $(A), (A)^*$ 中的诸 $[i]$ 与 $[i]^*$ 满足 $[j] \cap [k]^* = \varnothing$, 如果 $j+k < n$. 因而 $F_{\lambda\alpha}$ 与 $F_{\lambda\alpha}^*$ 也唯一地交于一点 $\{\alpha\}_0$, 且在该处单一相交. 由此知, $F_{\lambda\alpha} \times F_{\lambda\beta}$ 与 $F_{\lambda\alpha}^* \times F_{\lambda\beta}^*$ 也在 $\Omega_\alpha^n \times \Omega_{\beta-\alpha-1}^{n-\alpha-1}$ 中唯一地交于一点 $(\{\alpha\}_0, \{\beta-\alpha-1\}_0)$, 且在该处单一相交. 置 $\{\beta\}_0 = \{\alpha\}_0 \circ \{\beta-\alpha-1\}_0$, 则 $(\{\alpha\}_0, \{\beta\}_0)$ 是 F_λ 与 F_λ^* 的唯一交点, 在 T 之下与 $(\{\alpha\}_0, \{\beta-\alpha-1\}_0)$ 相对应. 由于 T 在该点双正则, 故由相交指数的双有理不变性 (参阅如 P. Samuel, *Méthodes d'algèbrce abstraite en géométric algébrique*, II §5. No. 4) 知, F_λ 与 F_λ^* 在 $\tilde{\Omega}_{\alpha,\beta}^n$ 中也在 $(\{\alpha\}_0, \{\beta-\alpha-1\}_0)$ 处单一相交, 即 (9) 式成立. 因而对偶有理分割定义中的条件 (iii) 也满足. 定理得证.

由于不论序列 (A) 的选择如何, 相应的 F_λ 恒属于同一个代数等价系 $\mathscr{G}_\lambda$, 因之由 §1 的定理, $\tilde{\Omega}_{\alpha,\beta}^n$ 的任一代数等价系可唯一表达成这些相同维数的代数等价系 $\mathscr{G}_\lambda$ 之和. 特别当 $\alpha = 0$ 时, 在 [10] 中引入的 d 维代数簇上的示性系有一组基本系, 可使之与符号 $\lambda \leftrightarrow \begin{bmatrix} a_i & & \\ a_0 & \cdots & a_d \end{bmatrix}$ 相对应. 而一般的示性系都是这些基本系的线性和.

参考文献

[1] Baldassari M. *Algebraic Varieties*. Berlin: Springer Verlag, 1956.

[2] W. L. Chow (周炜良). Algebraic varieties with rational dissections. *Proc. Nat. Acad. Sci.*, 1956, 42: 116-119.

[3] Ehresmann C.. Sur la topologie de certains espaces homogènes. *Ann. of Math.*, 1934, 35: 396-443.

[4] Hodge W. V. D. and Pedoe D. *Methods of Algebraic Geometry*, II. Cambridge, 1952.

[5] Igusa J. On the Picard variety attached to algebraic varieties. *Amer. J. Math.*, 1952, 74: 1-22.

[6] ———. Samuel P. Rational equivalence of arbitrary cycles. *Amer. J. Math.*, 1956, 78: 383-400.

[7] Relations d'équivalence en géométrie algébrique. *Proc. Intern. Cong. Mathematicians*, 1958. Cambridge, 1960: 470-487.

[8] Weil A. *Foundations of Algebraic Geometry*. New York, 1946.

[9] Weil A. Sur les critéres d'équivalence en géométrie algébrique. *Math. Ann.*, 1954, 128: 95-127.

[10] 吴文俊. 代数簇上的陈省身示性系. 数学进展, 1965, 8: 391-397.

代数簇上的陈省身示性系*

任一复流形 M 有一组陈省身示性类 $C^{2i} \in H^{2i}(M)$. 如果 M 同时是一个没有奇点的代数簇, 则 Gamkrelidze[2] 与陈省身[1] 曾证明了 C^{2i} 都是代数的, 即 C^{2i} 中有上闭链对偶于以 M 的代数子簇为代表的下闭链. 这自然引起了如何从代数几何方法对代数簇引入与陈省身示性类相仿的概念的问题. 在 [3] 中, Grothendieck(以及 Washnitzer 在 [6] 中) 引进了代数簇的陈省身示性系, 但须假定代数簇是没有奇点的. 这个没有奇点的限制似乎是难以避免的, 因为: 第一, 他的方法须借助于流形的切丛, 但在有奇点时, 切丛则无从定义; 第二, 他的方法须用到代数流形上的 (有理等价) 交截环, 而在有奇点时, 这个环也没有圆满的定义. 本文的目的在于应用另一种方法对任意代数簇引入陈省身示性系. 另外一个与 [3, 6] 不同之点是: 我们所定义的示性系是代数等价系, 而非如 [3, 6] 那样是有理等价系. 这对代数几何的问题来说, 似乎更适当些.

以下引用的概念与符号, 若非另有声明, 都依据 [4] 一书. 但 Cayley 点, Cayley 形式等名称都将改为周炜良点、周炜良形式等. 又, 我们所考虑的代数簇的基本域 K 总假定是代数闭且特征 0 的, 这些都不再明述.

§1. 设 M 是一个 K 上的不可约 b 维代数簇, M^0 是 M 的一个固定子簇, 含有 M 的全部奇点. 相对于 M^0 而言, M 的一个 c 维复合子簇 (multiplicative variety) $\mathbb{V}_c$ 将称为不可略的, 如果 $\mathbb{V}_c$ 的每一不可约成分 (components) 都不在 M^0 上. 由 M 上一个不可约子簇 V_c 所定, 以 1 为系数的复合簇将记作 $\mathbb{V}_c$, 称为 V_c 所定的单簇. 一个形式和 $\mathbb{V}_c = \sum a_i \mathbb{V}_c^{(i)}$ (这里 a_i 是任意整数, $\mathbb{V}_c^{(i)}$ 都是 M 上的单簇) 将称为一个 M 上的 c 维拟簇 (virtual variety), 以 $\mathbb{V}_c^{(i)}$ 为其成分. M 上的拟簇将称为不可略的, 如果它的每一成分都是不可略的. 这些 c 维不可略拟簇自然成一加法群, 与 [4]XII§§9, 10 同样, 在不可略拟簇之间可引进代数等价关系如下: 两个 M 上不可略复合子簇 $\mathbb{V}_c$ 与 $\mathbb{V}_c'$ 将称为不可略狭义等价, 记作 $\mathbb{V}_c \equiv \mathbb{V}_c' \mathrm{rel.} M^0$, 如果 M 上有一组不可略复合子簇 $\mathbb{V}_c^{(1)} = \mathbb{V}_c, V_c^{(2)}, \cdots, \mathbb{V}_c^{(k)} = \mathbb{V}_c'$, 使得对任意 $i = 1, 2, \cdots, k-1, \mathbb{V}_a^{(i)}$ 与 $\mathbb{V}_a^{(i+1)}$ 同属于 M 上一不可约子簇系. 如果 M 上有不可略复合子簇 $\mathbb{V}_c''$ 使 $\mathbb{V}_c + \mathbb{V}_c'' \equiv \mathbb{V}_c' + \mathbb{V}_c'' \mathrm{rel.} M^0$, 则称 $\mathbb{V}_c$ 与 $\mathbb{V}_c'$ 不可略广义等价, 记作 $\mathbb{V}_c \equiv \mathbb{V}_c' \mathrm{rel.} M^0$. M 上两个不可略拟簇 $\mathbb{V}_c, \mathbb{V}_c'$ 将称为不可略拟等价, 记作 $\mathbb{V}_c \sim \mathbb{V}_c' \mathrm{rel.} M^0$, 如果它们各可表作 $\mathbb{V}_c = \mathbb{V}_c' - \mathbb{V}_c''$, $\mathbb{V}_c = \mathbb{V}_c' - \mathbb{V}_c''$, 其中 $\mathbb{V}_c'$ 等都是不可略复合子簇, 而有 $\mathbb{V}_c' + \mathbb{V}_c'' \equiv \mathbb{V}_c'' + \mathbb{V}_c' \mathrm{rel.} M^0$. 与 [4]XII§§9, 10 同样可

---

*本文摘自《数学进展》, 第 8 卷第 4 期, 1965 年 11 月.

证 ≡ ≡, ∼rel. M^0 都是等价关系. M 上不可略拟簇在 ∼ rel.M^0 之下的等价类将称为 M 上相对于 M^0 而言的不可略代数等价系. 由一个不可略拟簇 $\widehat{V}_c$ 所定的不可略代数等价系将记作 $\{\widehat{V}_c\}^0$. 这些系在自然加法之下成一加法群, 记作 $\mathscr{A}_c(M/M^0)$, 而由通常意义下 c 维的代数等价系 (system of equivalence, [4] XII) 所构成的加法群将记为 $\mathscr{A}_c(M)$. 一个不可略代数等价系中的拟簇自然都属于同一个代数等价系, 因之有一个自然同态:

$$j_*: \mathscr{A}_c(M/M^0) \to \mathscr{A}_c(M).$$

§2. 今设 $\tilde{M}$ 是 $S_{\tilde{n}}$ 而 M 是 S_n 中的不可约代数簇, 维数都是 b, $\tilde{M}^0$ 与 M^0 各是 $\tilde{M}$ 与 M 的子簇, 各含有 $\tilde{M}$ 与 M 的全部奇点, 而 T 是 $\tilde{M}$ 到 M 的一个双有理变换, 具有以下几个性质:

1° T 在 $\tilde{M}$ 上全部有定义;
2° $\tilde{x} \in \tilde{M}^0$ 时, $T(\tilde{x}) \in M^0$, 而 $\tilde{x} \notin \tilde{M}^0$ 时 $T(\tilde{x})$ 也 $\notin M^0$;
3° T 在 $\tilde{M} - \tilde{M}^0$ 上的对应是一对一的.

命题 1 在上述假定下, 双有理变换 T 自然引出一同态:

$$T_*: \quad \mathscr{A}_c(\tilde{M}/\tilde{M}^0) \to \mathscr{A}_c(M/M^0).$$

证. 显然在 T 之下, $\tilde{M}$ 上一个不可略的 c 维不可约子簇 $\tilde{V}_c$ 对应为 M 上一个不可略的同维不可约子簇, 记作 $T_\#\tilde{V}_c$, 因而 T 也自然地把 $\tilde{M}$ 上不可略 c 维拟簇 $\widehat{V}_c$ 对应为 M 上的不可略 c 维拟簇, 记作 $T_\#\widehat{V}_c$. 于是要证明本命题, 只须证明下一陈述:

如果 $\tilde{M}$ 上不可略的 c 维不可约子簇 $\tilde{U}_c$ 与 $\tilde{V}_c$ 属于同一 $\tilde{M}$ 上子簇所成某不可约代数系 $\tilde{\mathscr{C}}$, 则 $T_\#\tilde{U}_c = U_c$ 与 $T_\#\tilde{V}_c = V_c$ 也同属于一 M 上子簇所成的某不可约代数系.

为此, 设 $\tilde{U}_c$ 与 $\tilde{V}_c$ 的公共次数 (order) 为 $\tilde{g}$. 命 $F(z; u_0, \cdots, u_c)$ 为对每组不定量 $(u_{i0}, \cdots, u_{in})(i = 0, \cdots, c)$ 都是 $\tilde{g}$ 次齐次的多项式, 以不定量 $z_0, \cdots, z_D$ 为系数. 于是 $F(z'; u_0, \cdots, u_c)$ 作为 $\tilde{M}$ 上一个 c 维 $\tilde{g}$ 次子簇的周炜良形式, 它的充要条件是 z' 需满足一组齐次代数方程 ([4] 页 58):

$$T_\omega(z') = 0 \quad (\omega = 1, 2, \cdots) \tag{1}$$

方程组 (1) 确定了 S_D 中的一个代数簇 $\tilde{I}$, 而对应于 $\tilde{\mathscr{C}}$ 中子簇的周炜良点 (z) 构成 $\tilde{I}$ 的一个不可约子簇 $\tilde{I}'$. 设由方程组

$$f_i(z) = 0 \quad (i = 1, 2, \cdots) \tag{2}$$

所定, 而 (1) 包含在 (2) 之内. 命 $S^0, \cdots, S^d$ 为 $d+1$ 个 $\tilde{n}+1$ 阶反对称矩阵, 其元素 $s^i_{jk}(k>j)$ 为独立的不定量, 则对于 $z \in \tilde{I}'$, 它所对应的子簇 $\tilde{V}(z) \in \tilde{\mathscr{C}}$ 由方程组

$$F(z; S^0\tilde{x}, \cdots, S^d\tilde{x}) = 0 \tag{3}$$

决定, 这里 $\tilde{x} = (\tilde{x}_0, \cdots, \tilde{x}_{\tilde{n}}) \in S_{\tilde{n}}$. 今设 T 的定义方程为 $(x = (x_0, \cdots, x_n) \in S_n)$

$$x_i g_j(\tilde{x}) = x_j g_i(\tilde{x}) \qquad i, j = 0, 1, \cdots, n. \tag{4}$$

从 (3), (4) 消去 $\tilde{x}$ 得结式组

$$R_k(z, x) = 0 \qquad (k = 1, 2, \cdots). \tag{5}$$

于是 (2) 与 (5) 确定了一个代数对应 C, 它的原簇 (object variety) 显然是 $\tilde{I}'$, 而象簇 (image variety) $N \subset M$, 试考察 $z' \in \tilde{I}'$ 的象簇 $V(z')$. 设 $x' \in V(z')$, 则 (z', x') 满足 (5), 故有 $\tilde{x}' \in S_{\tilde{\#}}$ 使 $(z', \tilde{x}')$ 满足 (3) 而 $(\tilde{x}', x')$ 满足 (4). 但 (4) 式意谓 $x' = T\tilde{x}'$ 而 (3) 式意谓 $\tilde{x}' \in \tilde{V}(z')$, 由此得 $V(z') = T_\# \tilde{V}(z')$. 因为 C 的原簇 $\tilde{I}'$ 不可约, 而每一 $z' \in \tilde{I}'$ 的象簇又是同一维数 c 的不可约代数簇, 故 C 是一个不可约的代数对应 ([5] 页 146 习题).

今对 $\tilde{I}'$ 的一个一般点 (generic point)ξ 命 $V(\xi) = T_\# \tilde{V}(\xi)$ 的周炜良点为 ζ, 于是一般点偶 (ζ, ξ) 确定了一个 $\tilde{I}'$ 与某一不可约代数簇 I' 间的不可约代数对应, 这里 I' 中的点都是 M 上 c 维而次数同于 $V(\zeta)$ 的子簇的周炜良点. 这个 I' 确定了 M 上一个由 c 维子簇所成的不可约代数系 $\mathscr{C}$. 由 [4]XI§6 定理 II, 在这一对应中, $z' \in \tilde{I}'$ 只有唯一的一个对应点, 即 $V(z')$ 的周炜良点. 特别是当 z' 分别是 $\tilde{U}_c$ 与 $\tilde{V}_c$ 的周炜良点时, 它们的对应点也就是 U_c 与 V_c 的周炜良点. 因而 U_c 与 V_c 属于 M 上同一不可约代数系 $\mathscr{C}$. 这证明了上面的陈述, 因而也证明了命题 1.

§3. 设 V_n 是 S_N 中一个不可约 n 维代数簇, 没有奇点, 因而依 [4]XII§7, 以下在 V_n 上可定义交截 (intersection).

今设 $\tilde{M}$ 是 V_n 上一个不可约 b 维子簇, $\tilde{M}^0$ 是 $\tilde{M}$ 在 $\tilde{M}$ 的一个固定子簇, 含有 $\tilde{M}$ 的全部奇点. 设 $\mathscr{C} \in \mathscr{A}_a(V_n)$, 依 [4]XII§10 定理 2 的证明, 在 $\mathscr{C}$ 中有拟簇 $\mathbb{O}_a = \sum \sigma_i \mathbb{O}_a^{(i)}$, 这里 $\mathbb{O}_a^{(i)}$ 是由 V_n 上一个不可约子簇 $U_a^{(i)}$ 所定的单簇, 使每一 $U_a^{(i)}$ 与 $\tilde{M}$ 简单相交 (intersect simply), 也与 $\tilde{M}^0$ 简单相交 (这时将称 $\mathbb{O}_a$ 与 $\tilde{M}, \tilde{M}^0$ 简单相交). 由此知 $\dim(U_a^{(i)} \wedge \tilde{M}^0) < \dim(U_a^{(i)} \wedge \tilde{M}) = a+b-n$, 因之 $U_a^{(i)} \wedge \tilde{M}$ 是 $\tilde{M}$ 上相对于 $\tilde{M}^0$ 而言的不可略子簇, 而 $\mathbb{O}_a \cdot \tilde{M}$ 是 $\mathbb{O}_a \cdot \tilde{M}$ 在 $\tilde{M}$ 上的不可略拟簇.

命题 2 设 $V_n, \tilde{M}, \tilde{M}^0$ 与 $\mathscr{C} \in A_a(V_n)$ 如上, 则对任一 $\mathbb{O}_a \in \mathscr{C}$, 这里 $\mathbb{O}_a$ 各与 $\tilde{M}, \tilde{M}^0$ 简单相交时, $\mathbb{O}_a \cdot \tilde{M}$ 属于 $\tilde{M}$ 上一个确定的不可略代数等价系, 记之为

$\mathscr{C} \cdot \tilde{M}$. 于是 $\mathscr{C} \to \mathscr{C} \cdot \tilde{M}$ 是一个同态

$$I_* : A_a(V_n) \to A_{a+b-n}(\tilde{M}/\tilde{M}^0).$$

证. 我们只须证明下述各点. 设 $Ⓤ_a$ 与 $Ⓥ_a \in \mathscr{C}$ 各与 $\tilde{M}, \tilde{M}^0$ 简单相交, 则

$$Ⓤ_a \cdot \tilde{M} \sim Ⓥ_a \cdot \tilde{M} \qquad \text{rel.} \tilde{M}^0. \tag{1}$$

为此先依 Severi 与 Van der Waerden 引进一 V_n 上复合子簇间的运算如下: 设 $Ⓐ_a = \sum \sigma_i Ⓐ_a^{(i)} (\sigma_i > 0)$ 是 V_n 上一个复合子簇, 其中 $Ⓐ_a^{(i)}$ 都是不可约子簇 $A_a^{(i)}$ 所定的单簇. 在 S_N 中取一与 $Ⓐ_a$ 无关即与每一 $A_a^{(i)}$ 都无关 (independent) 的 S_{N-n-1}, 作它与 $A_a^{(i)}$ 的联合 (join) $W_{N-n+a}^{(i)}$, 并记 $Ⓦ_{N-n+a} = \sum \sigma_i Ⓦ_{N-n+a}^{(i)}$, 称之为 $Ⓐ_a$ 与 S_{N-n-1} 的联合. 于是 $Ⓦ_{N-n+a} \cdot V_n = Ⓐ_a + Ⓐ_a^*$, 这里 $Ⓐ_a^*$ 是 V_n 上的一个复合子簇, 我们将记之为

$$Ⓐ_a^* = O(S_{N-n-1}) \cdot Ⓐ_a.$$

这个由 $Ⓐ_a$ 到 $Ⓐ_a^*$ 的对应导源于 Severi, 它有以下一些性质:

(i) 如果 $Ⓐ_a$ 与 $\tilde{M}, \tilde{M}^0$ 都简单相交, 则 $Ⓐ_a^*$ 也与 $\tilde{M}, \tilde{M}^0$ 简单相交 (依据 [4]XII§6 引理 V, VI).

(ii) 依次作 $Ⓐ_a^{**} = O(S'_{N-n-1}) \cdot Ⓐ_a^*, \cdots, Ⓐ_a^{(*')} = O(S_{N-n-1}^{(r-1)}) \cdot Ⓐ_a^{(*r-1)})$, 这里 S'_{N-n-1} 与 $Ⓐ_a^*$ 无关; 余类推. 于是在 r 充分大时, $Ⓐ_a^{(*')}$ 与 $\tilde{M}, \tilde{M}_0$ 都简单相交 (依据 [4]XII§6 引理 IV – VI, 参阅 [4]XII§10 定理 II 的证明).

(iii) 设 $Ⓐ_a$ 等同 (ii), 则 $Ⓐ_a$ 与 $Ⓐ_a^{**}$ 属于同一不可约代数系.

盖取一一般 (generic) 的 $\bar{S}_{N-n-1}$, 则 $\bar{S}_{N-n-1}$ 可特定化为 S_{N-n-1} 或 S'_{N-n-1}, 因而 $\bar{S}_{N-n-1}$ 与 $Ⓐ_a^*$ 的联合 $Ⓦ_{N-n+a}$ 可特定化为 S_{N-n-1} 与 $Ⓐ_a^*$, 即 S_{N-n-1} 与 $Ⓐ_a$ 的联合 $Ⓦ_{N-n+a}$, 也可特定化为 S'_{N-n-1} 与 $Ⓐ_a^*$ 的联合 $Ⓦ_{N-n+a}^*$ (见 [4]XII§8 定理 I 的证明), 故由 [4]XII§8 定理 II, 有特定化 $\bar{Ⓦ}_{N-n+a} \cdot V_n \to Ⓦ_{N-n+a} \cdot V_n$ 与 $\bar{Ⓦ}_{N-n+a} \cdot V_n \to Ⓦ_{N-n+a}^* \cdot V_n$. 命 $\bar{Ⓦ}_{N-n+a} \cdot V_n = \bar{Ⓐ}_a^* + \bar{Ⓐ}_a^*$, 即有特定化 $\bar{Ⓐ}_a^* \to Ⓐ_a$ 与 $\bar{Ⓐ}_a^* \to Ⓐ_a^{**}$, 则 $Ⓐ_a$ 与 $Ⓐ_a^{**}$ 属于同一以 $\bar{Ⓐ}_a^*$ 为一般元素的不可约代数系, 如所欲证.

(iv) 设 $Ⓐ_a$ 与 $\bar{Ⓐ}_a$ 都是 V_n 上的复合子簇, 而 $Ⓐ_a$ 是 $\bar{Ⓐ}_a$ 的一个特定化. 取 S_{N-n-1} 与 $Ⓐ_a$ 无关, 也与 $\bar{Ⓐ}_a$ 无关, 并置 $Ⓐ_a^* = O(S_{N-n-1})Ⓐ_a, \bar{Ⓐ}_a^* = O(S_{N-n-1})\bar{Ⓐ}_a$, 则在特定化 $\bar{Ⓐ}_a \to Ⓐ_a$ 下有特定化 $\bar{Ⓐ}_a^* \to Ⓐ_a^*$.

盖设 $Ⓐ_a, \bar{Ⓐ}_a$ 与 S_{N-n-1} 的联合各为 $Ⓦ_{N-n+a}$ 与 $\bar{Ⓦ}_{N-n+a}$, 则在特定化 $\bar{Ⓐ}_a \to Ⓐ_a$ 下有特定化 $\bar{Ⓦ}_{N-n+a} \to Ⓦ_{N-n+a}$, 而由 [4]XII§8 定理 II, 也有特定化 $\bar{Ⓦ}_{N-n+a} \cdot V_n \to Ⓦ_{N-n+a} \cdot V_n$, 因而有特定化 $\bar{Ⓐ}_a^* \to Ⓐ_a^*$, 如所欲证.

今试证 (1) 式如下：为此记

$$\mathcal{U}_a = \mathcal{U}'_a - \mathcal{U}''_a, \quad \mathcal{V}_a = \mathcal{V}'_a - \mathcal{V}''_a,$$

这里 $\mathcal{U}'_a$ 等都是复合子簇，各与 $\tilde{M}$ 及 $\tilde{M}^0$ 简单相交．由于 $\mathcal{U}_a, \mathcal{V}_a$ 都 $\in \mathscr{C}$，故在 V_n 上有复合子簇 $\mathcal{A}_a$ 使得在 V_n 上有

$$\mathcal{A}_a + \mathcal{U}'_a + \mathcal{V}''_a \equiv \mathcal{A}_a + \mathcal{U}''_a + \mathcal{V}'_a. \tag{2}$$

我们不妨假定 $\mathcal{A}_a$ 与 $\tilde{M}$ 及 $\tilde{M}^0$ 简单相交．盖设不然，可取 S_N 中与 $\mathcal{A}_a$ 无关的 S_{N-n-1} 而作它与 $\mathcal{A}_a$ 的联合 $\mathcal{W}_{N-n+a}$，于是有 $\mathcal{W}_{N-n+a} \cdot V_n = \mathcal{A}_a + \mathcal{A}_a^*$，而由 (2) 得

$$\mathcal{W}_{N-n+a} \cdot V_n + \mathcal{U}'_a + \mathcal{V}''_a \equiv \mathcal{W}_{N-n+a} \cdot V_n + \mathcal{U}''_a + \mathcal{V}'_a \tag{3}$$

(在 V_n 上)．在 S_N 中取——般投影变换 τ，并作 $\mathcal{W}_{N-n+a}$ 在 τ 下的变换簇 $\mathcal{W}^r_{N-n+a}$，则因取特定化 $\tau \to I(I = $ 互同投影变换$)$ 时有特定化 $\mathcal{W}^\tau_{N-n+a} \cdot V_n \to \mathcal{W}_{N-n+n} \cdot V_n$，故从 (3) 得 (在 V_n 上):

$$\mathcal{W}^\tau_{N-n+a} \cdot V_n + \mathcal{U}'_a + \mathcal{V}''_a \equiv \mathcal{W}^\tau_{N-n+a} \cdot V_n + \mathcal{U}''_a + \mathcal{V}'_d.$$

由 [4]XII§3 定理 I，$\mathcal{W}^\tau_{N-n+a} \cdot V_n$ 与 $\tilde{M}, \tilde{M}^0$ 都简单相交，因而可取为 (2) 中的 $\mathcal{A}_a$. 今在 (2) 中置

$$\left.\begin{array}{l}\mathcal{X}_a = \mathcal{A}_a + \mathcal{U}'_a + \mathcal{V}''_a, \\ \mathcal{Y}_a = \mathcal{A}_a + \mathcal{U}''_a + \mathcal{V}'_a,\end{array}\right\} \tag{4}$$

这里 $\mathcal{X}_a, \mathcal{Y}_a$ 各与 $\tilde{M}, \tilde{M}^0$ 简单相交，于是有一 V_n 上复合子簇的序列

$$\mathcal{V}^{(1)}_a = \mathcal{X}_a, \mathcal{V}^{(2)}_a, \cdots, \mathcal{V}^{(k)}_a = \mathcal{Y}_a,$$

使 $\mathcal{V}^{(i)}_a$ 与 $\mathcal{V}^{(i+1)}_a$ 属于同一不可约代数系 $(i = 1, 2, \cdots, k-1)$．命 $\mathcal{U}^{(i)}_a$ 为它的一个一般元素，取一与诸 $\mathcal{V}^{(i)}_a, \mathcal{U}^{(i)}_a$ 都无关的 S_{N-n-1} 以作 $O(S_{N-n-1})\mathcal{V}^{(i)}_a = \mathcal{V}^{(i)*}_a, O(S_{N-n-1})\mathcal{U}^{(i)}_a = \mathcal{U}^{(i)*}_a$，又取一与诸 $\mathcal{V}^{(i)*}_a, \mathcal{U}^{(i)*}_a$ 都无关的 S'_{N-n-1} 以作 $O(S'_{N-n-1}) \cdot \mathcal{V}^{(i)*}_a = V^{(i)**}_a, O(S'_{N-n-1})\mathcal{U}^{(i)*}_a = \mathcal{U}^{(i)**}_a$，等．依次类推，得一序列

$$\mathcal{V}^{(1)(*')}_a = \mathcal{X}^{(*')}_a, \mathcal{V}^{(2)(*')}_a, \cdots, \mathcal{V}^{(k-1)(*')}_a, \mathcal{V}^{(k)(*')}_a = \mathcal{Y}^{(*')}_a. \tag{5}$$

由 (iv)，序列中每两相继元素都属于同一不可约代数系．由 (ii)，当取 r 充分大时，序列 (5) 中每一元素都与 $\tilde{M}, \tilde{M}^0$ 简单相交，因而依 [4]XII§8 定理 II，下述序列:

$$\mathcal{V}^{(1)(*')}_a \cdot \tilde{M}, \mathcal{V}^{(1)(*')}_a \cdot \tilde{M}, \mathcal{V}^{(2)(*')}_a \cdot \tilde{M}, \cdots, \mathcal{V}^{(k)(*')}_a \cdot \tilde{M} \tag{6}$$

中每一元素都是 $\tilde{M}$ 上不可略复合子簇, 且每两相继元素都属于 $\tilde{M}$ 上同一不可略子簇系, 或

$$\widetilde{X}_a^{(*')} \cdot \tilde{M} \equiv \widetilde{Y}_a^{(*')} \cdot \tilde{M} \text{ rel.} \tilde{M}^0. \tag{7}$$

特别取 r 为偶数, 则由 (i), (iii) 以及 [4]XII§8 定理 II, 又有

$$\left.\begin{array}{l}\widetilde{X}_a \cdot \tilde{M} \equiv \widetilde{X}_a^{(*')} \cdot \tilde{M} \text{ rel.} \tilde{M}^0, \\ \widetilde{Y}_a \cdot \tilde{M} \equiv \widetilde{Y}_a^{(*')} \cdot \tilde{M} \text{ rel.} \tilde{M}^0.\end{array}\right\} \tag{8}$$

由 (7), (8) 即得

$$\widetilde{X}_a \cdot \tilde{M} \equiv \widetilde{Y}_a \cdot \tilde{M} \text{ rel.} \tilde{M}^0.$$

依 $\widetilde{X}_a, \widetilde{Y}_a$ 的定义, 上式即 (1) 式, 命题得证.

§4. 对于 K 上的 n 维投影空间 $S_n(K)$, 命 $\tilde{\Omega}_{d,n}$ 为所有偶 (z, L) 的集体, 这里 $z \in S_n(\tilde{K})$, 而 L 是 $S_n(\tilde{K})$ 中任一含有 z 的 d 维投影子空间, $\tilde{K}$ 则是 K 的任一扩域. 于是 $\tilde{\Omega}_{d,n}$ 是一代数簇 (为一复合格拉斯曼流形), 没有奇点, 且可自然地实现于某一 $S_N(K)$ 中. 以后将假定如此.

今设 M 是 $S_n(K)$ 中域 K 上的一个 d 维不可约代数簇. 命 ξ 是 M 的任一一般点, L_ξ 是 M 在 ξ 的切面, 于是 (ξ, L_ξ) 作为 $\tilde{\Omega}_{d,n}$ 的点确定了 $\tilde{\Omega}_{d,n}$ 的一个子簇 $\tilde{M}$, 以之为对 K 的一般点. 我们有

命题 3 $\tilde{M}$ 是 $\tilde{\Omega}_{d,n}$ 的 d 维子簇, 与一般点 $\xi \in M$ 的选择无关, 且对 $(z, L) \in \tilde{M}$ 命 $T(z, L) = z \in M$, 又命 $\tilde{M}^0$ 为由 $\tilde{M}$ 上所有奇点以及所有使 z 为 M 奇点的点 (z, L) 所成的子簇, 而 M^0 为 $\tilde{M}^0$ 在 T 下的象, 则 T 是 $\tilde{M}$ 到 M 上的一个双有理变换, 满足 §2 中的性质 $1° - 3°$.

证. 在 $\xi = (\xi_0, \cdots, \xi_n)$ 中不妨设 $\xi_0 = 1, \xi_1, \cdots, \xi_d$ 对 K 代数无关, 于是在 M 的方程组中有 $n - d$ 个不可约方程

$$g_i(x_0, \cdots, x_d, x_{d+i}) = 0, \quad i = 1, \cdots, n-d,$$

这里 $g_i \in K[x]$. 由 [4]X§14 定理 I 知 M 在 ξ 的切面 L_ξ 由方程组

$$\sum_{j=1}^n \frac{\partial g_i}{\partial \xi_i} x_j = 0, \quad i = 1, \cdots, n-d$$

决定, 因而 L_ξ 的 Plücker 坐标即行列式 $\left(\frac{\partial g_i}{\partial \xi_j}\right)$ 的诸 $n-d$ 阶子行列式. 由于 $\frac{\partial g_i}{\partial \xi_j}$ 是 $\xi_1, \cdots, \xi_n$ 的多项式, 因之代数依赖于 $\xi_1, \cdots, \xi_d$, 这些 Plücker 坐标也代数依赖于 $\xi_1, \cdots, \xi_d$. 由此知 (ξ, L_ξ) 的坐标的超越度 $\leqslant d$, 另一方面又显然 $\geqslant d$, 故超越度 $= d$. 从而知 $\tilde{M}$ 为 d 维代数簇.

其次, 设 ξ' 是 M 的另一一般点, $L_{\xi'}$ 是 M 在 ξ' 的切面, 于是要证明 $\tilde{M}$ 与 ξ 的选择无关, 只须证: 在对 K 的特定化 $\xi \to \xi'$ 下有特定化 $(\xi, L_\xi) \to (\xi', L_{\xi'})$ 即可. 为此, 设 M 的方程组为

$$f_i(x) = 0 \quad (i = 1, 2, \cdots),$$

并考虑超平面

$$H_i : \sum \frac{\partial f_i}{\partial \xi_j} \cdot x_j = 0,$$

与

$$H_i' : \sum \frac{\partial f_i}{\partial \xi_j'} \cdot x_j = 0.$$

于是 L_ξ(与 $L_{\xi'}$) 恰是所有 H_i(与 H_i'), $i = 1, 2, \cdots$ 的公共交面. 命 L_ξ 的 Plücker 坐标为 $p_{j_0 \cdots j_d}$. 设在特定化 $\xi \to \xi'$ 下, L_ξ 的特定化为一以 $p'_{j_0 \cdots j_d}$ 为 Plücker 坐标的 d 维投影空间 L'. 依 [4]VII§5 定理 I, $L_\xi \subset H_i$ 的充要条件可用 $p_{j_0 \cdots j_d}$ 与 $\frac{\partial f_i}{\partial \xi_j}$ 的一组齐次多项式 $= 0$ 的关系来表示. 在特定化 $(\xi, L_\xi) \to (\xi', L')$ 下, 这些多项式关系变为 $p'_{j_0 \cdots j_d}$ 与 $\frac{\partial f_i}{\partial \xi_j'}$ 间的同样一组关系, 亦即应有 $L' \subset H_i'$. 因之 L' 恰是诸 H_i' 的交面而有 $L' = L_{\xi'}, (\xi, L_\xi) \to (\xi', L_{\xi'})$, 如所欲证.

命题的最后一部分是显然的.

§5. 设 M 如 §4. 根据命题 3, 从 M 可唯一地作一 $\tilde{\Omega}_{d,n}$ 中的子簇 $\tilde{M}$, 并由命题 1 有同态

$$T_* : \quad \mathcal{A}_c(\tilde{M}/\tilde{M}^0) \to \mathcal{A}_c(M/M^0).$$

由于 $\tilde{\Omega}_{d,n}$ 没有奇点, 故由命题 2 又有同态

$$I_* : \quad \mathcal{A}_c(\tilde{\Omega}_{d,n}) \to \mathcal{A}_{c+d-r}(\tilde{M}/\tilde{M}^0),$$

其中 $r = (d+1)(n-d) + d$ 是 $\tilde{\Omega}_{d,n}$ 的维数. 此外又有自然同态

$$j_* : \quad \mathcal{A}_c(M/M^0) \to \mathcal{A}_c(M).$$

定义 对任意 $\mathscr{C} \in \mathcal{A}_{r-s}(\tilde{\Omega}_{d,n}), 0 \leqslant s \leqslant d,$

$$j_* T_* I_*(\mathscr{C}) \in \mathcal{A}_{d-s}(M)$$

将称为 S_n 中代数簇 M 对应于 $\mathscr{C}$ 的示性系, 记作 $\mathscr{C}_\mathscr{I}(M)$.

在 M 的示性系中, 有一些特别重要. 为此, 命

$$S_{n-d-2+t} \subset S_{n-s+t}$$

为 S_n 中任两一般的 (generic) 投影子空间, 这里 $0 \leqslant t \leqslant s \leqslant d$. 于是满足 $z \in S_{n-s+t}$ 与 $\dim(S_d \cap S_{n-d-2+t}) \geqslant t-1$ 的一切偶 $(z, S_d) \in \tilde{\Omega}_{d,n}$ 成一 $r-s$ 维不可约子簇 $V_{r-s}^{(t)}$ (依 Ehresmann 记号, 这是一个 Schubert 子簇 $\begin{bmatrix} n-s-t \\ n-k-1, \cdots, n-k-2 \\ +t, n-k+t, \cdots, n \end{bmatrix}$). 由 $V_{r-s}^{(t)}$ 所定的代数等价系将记作 $v_{r-s}^{(t)} \in \mathcal{A}_{r-s}(\tilde{\Omega}_{d,n})$. 置

$$v_{r-s} = \sum_{t=0}^{s}(-1)^t \binom{d-t-1}{d-s-1} v_{r-s}^{(t)} \in \mathcal{A}_{r-s}(\tilde{\Omega}_{d,n}), \tag{1}$$

$$\Gamma_{d-s}^{(t)}(M) = j_* T_* I_*(v_{r-s}^{(t)}) \in \mathcal{A}_{d-s}(M), \tag{2}$$

与

$$\mathscr{C}_{d-s}(M) = j_* T_* I_*(v_{r-s}) \in \mathcal{A}_{d-s}(M). \tag{3}$$

定义 $\Gamma_{d-s}^{(t)}(M)$ 与 $\mathscr{C}_{d-s}(M)$ 分别称为 M 的 Гамкрелидзе 与陈省身示性系. 这些示性系间有下述关系:

$$\mathscr{S}_{d-s}(M) = \sum_{t=0}^{s}(-1)^t \binom{d-t-1}{d-s-1} \Gamma_{d-s}^{(t)}(M). \tag{4}$$

这一公式称为 Гамкрелидзе 公式.

这些名称的引入基于下述定理: 我们知道基本域 K 是复数域时, d 维代数簇 M 作为复投影空间的子空间可定义上下同调群 $H^k(M)$ 与 $H_k(M)$, 而每一 M 上的拟簇 $\mathbb{V}_c$ 自然代表一 $2c$ 维下闭链. 在 M 没有奇点时, M 是 $2d$ 维自然定向闭流形, 因而有对偶同构

$$\mathscr{D}: \quad H^{2s}(M) \approx H_{2d-2s}(M).$$

作为复流形, M 上并有一组陈省身示性类:

$$\mathscr{C}^{2s}(M) \in H^{2s}(M), 0 \leqslant s \leqslant d.$$

我们知道, 这时每一 M 上 k 维代数等价系唯一地确定了一个 $H_{2k}(M)$ 中的下同调类, 即有同态 [1]

$$\lambda_*: \quad \mathcal{A}_k(M) \to H_{2k}(M).$$

于是有

---

1) 这一事实的证明, 可参阅 J.Igusa, On the Picard varieties attached to algebraic varieties. Amer.J. Math., 1952, 74: 1—22, §I.

定理 设基本域 K 是复数域, 而 M 是 S_n 中没有奇点的 d 维代数簇, 则有

$$\lambda_* \mathscr{C}_{d-s}(M) = \mathscr{DC}^{2s}(M). \tag{5}$$

证. 仍用前面的记号. 在定理假设之下, $\tilde{\Omega}_{d,n}, M, \tilde{M}$ 都是复流形而 $(z, S_d) \to z$ 引出一个 $\tilde{M}$ 到 M 上的复同构 φ. 如 [2], 选择 S_n 中投影子空间 $S_{n-d-2+t} \subset S_{n-s+t}$ 与 M 在一般位置以定义复流形 $V_{r-s}^{(t)}$, 则由 [2], 有

$$C_{2d-2s} = \sum_{t=0}^{s} (-1)^t \binom{d-t-1}{d-s-1} \varphi_*(V_{r-s}^{(t)} x \tilde{M}) \in \mathscr{DC}^{2s}(M), \tag{6}$$

这里 x 表 (拓扑) 交截. 另外, 由于对 $\tilde{\Omega}_{d,n}$ 中任一代数等价系 $\mathscr{C}$ 显然有

$$\lambda_* T_* I_*(\mathscr{C}) = \varphi_*(\lambda_* \mathscr{C} x \tilde{M}), \tag{7}$$

而 j_λ 在 M 无奇点时显然互同, 且 $V_{r-s}^{(t)}$ 显然属于代数等价系 $v_{r-s}^{(t)}$, 故从 (4), (6), (7) 即得 (5) 式, 如所欲证.

参考文献

[1] S.Chern(陈省身). On the characteristic classes of complex sphere bundles and algebraic varieties. *Amer. J. Math.,* 1953, 75: 565-597.

[2] Гамкрелидзе Р. В.. Циклы черна комплексных алгебраических многообразий, Изв.Ак.Наук СССР. Сер., МаТем, 1956, 20: 685-706.

[3] Grothendieck A.. La thérie des classes de Chern. *Bull. Soc. Math. France,* 1958, 86: 137-154.

[4] Hodge W. V. D., Pedoe D.. *Methods of Algebraic Geometry.* Cambridge, vol. Ⅰ, 1947; vol. Ⅱ, 1952.

[5] Van der Waerden. *Einfuhrung in die Algebraischen Geometrie.* Berlin, 1945.

[6] Washintzer G.. The characteristic classes of an algebraic fiber bundle Ⅱ. *Proc. Nat. Aced. Sci.,* 1955, 42: 433-436.

Chern Classes on Algebraic Varieties with Arbitrary Singularities*

In [2] the author introduced the notion of Chern classes for projective algebraic varieties with arbitrary singularities. Unlike the intricate method due to MacPherson et al (e.g.[1]), our method is simple and direct and may be briefly described as follows.

Let V_d be an irreducible algebraic variety in the projective space S_n on $\underline{C}$. Take a generic point P of V_d with tangent space L_P. The pair (P, L_P) will determine a subvariety $\tilde{v}_d$ of the composite Grassmannian $\tilde{\Omega}^n_{o,d}$ of pairs $\{S_O, S_d\}$ with $S_O \subset S_d \subset S_n$. Let $A^*(M) = \sum A^r(M)$ be the group of algebraic equivalence classes for an arbitrary variety M with $r =$ codimension. In the case of $M = \tilde{\Omega}^n_{o,d}$ which has no singularity, A^* is a ring under intersections and the basis in A are in correspondence with the Ehresmann symbols

$$E = \begin{bmatrix} n - a_i \\ n - a_d, \cdots, n - a_i, \cdots, n - a_o \end{bmatrix}$$

with

$$0 \leqslant a_0 < \cdots < a_i < \cdots < a_d \leqslant n, \quad r = \sum_{j=0}^{d}(a_j - j) + i.$$

To each such element $E \in A^r$ the intersection with V_d will give an element $\tilde{E} \in A^r(\tilde{v}_d)$ and thus an element $E(v_d) \in A^r(v_d)$, to be defined as the (generalized) characteristic class of V_d corresponding to the symbol E. Of particular interest are the elements

$$C^S(v_d) = \sum_{t=0}^{S}(-1)^t \binom{d-t-1}{d-s-1} E^s_t(v_d),$$

in which $E^s_t = \begin{bmatrix} n-s+t \\ n-d-1, \cdots, n-d-2+t, n-d+t, \cdots, n \end{bmatrix}.$

Let

$$h: A^r(V_d) \longrightarrow H_{2d-2r}(V_d, \underline{C})$$

---

* 本文原载 Kohn J. J., Remmert R., Lu Q.K., Siu Y.T. (eds). *Several Complex Variables.* Boston: Birkhäuser, 1984: 247-249.

be the natural morphism and in the case of non-singular V_d let

$$D : H_{2d-2r}(V_d, \underline{C}) \longrightarrow H^{2r}(v_d, \underline{C})$$

be the Poincaré duality, then by Gamkrelidze, $DhC^s(V_d)$ are just the usual Chern classes C_s of V_d.

For $s = 0$ the classes reduce to numbers which will be called (generalized) Chern numbers of V_d. These are just the enumerative characters of V_d in the sense of Severi. Under certain assumptions we have

The Hilbert polynomial of an arbitrary algebraic variety

is completely determined by its generalized Chern numbers.

The explicit expressions can then be determined in an elementary manner which generalizes the Hirzebruch formula for arithmetic genera in the non-singular case.

References

[1] R.D. MacPherson. Chern classes for singular algebraic varieties. Ann. of Math. 1974, 100: 423-432.

[2] W.-t. Wu. Shuxue Jinzhan, 1965, 8: 395-409.

On Chern Numbers of Algebraic Varieties with Arbitrary Singularities*

Abstract In 1965 the author introduced the notion of Chern classes for an algebraic variety with arbitrary singularities. Based on this definition the well-known Miyaoka-Yao inequalities have been proved and extended by quite simple direct computations.

In 1977 Miyaoka and Yau (see [M] and [Y]) have proved independently a remarkable inequality about Chern numbers of a SMOOTH algebraic surface S, viz.

$$c_1^2(S) <= 3 * c_2(S). \tag{MY}$$

Some results and conjectures of similar nature have also been anounced for high dimensional algebraic varieties (see e.g. [T], [Y]). Their considerations are all restricted to algebraic varieties without any singularities since tools for complex manifolds were used throughout. Now in 1977 MacPherson [MP] has introduced the notion of Chern classes for any algebraic variety with arbitrary singularities. It is natural to ask whether the above inequality remains true for this general case for which the present author is quite ignorant of the present status. On the other hand early in 1965 the present author has already generalized the notion of Chern classes to arbitrary algebraic varieties in an entirely different way from that of MacPherson et al, cf. [WU1—3]. It turns out that the formula (MY) and its alike can be easily dealt with by our treatment for varieties with singularities. This will be the main theme of the present paper. Other applications of our method will be dealt with later.

We use in this paper notations which, being readily done by computer-printing, are somewhat different from the usually adopted ones. For the convenience of the reader a comparison between these two types of some of these notations used are tabulated below:

---

* 本文原载 *Acta Mathematica Sinica*, New Series, 1987, 3(3): 221-236.

new notations	usual notations	explanations
$-<$	$\subset, \in$	"is contained in" or "belongs to"
$>-$	$\supset$	"contains"
$<=$	$\leq$	"less than or equal to"
$>=$	$\geq$	"greater than or equal to"
$<>$	$\neq$	"not equal to"
$*$		"multiply by"
$\char`^$		"to the power of" or "intersects with"
$(n//m)$	$\binom{n}{m}$	binomial coefficient

Sect 1. The composite Grassmann variety

Let us recall first some fundamental facts about composite Grassmann variety due to Ehresmann et al (see e.g. [EH], [HP], [CHOW], and [WU2]), which is at the basis of our treatment. Note that we are working in the complex domain so the modifier "complex" will often be omitted.

Consider thus a projective space CPn of dimension n. The linear subspaces of dimension k will be denoted by $[k], [k]', Sk, S'k$, etc. For fixed integers p, q with $0 < p < q < n$ the totality of pairs $([p], [q])$ with $[p]-<[q]-<[n]=CPn$ will be denoted henceforth as $GR(n; p, q)$. It is a special kind of composite Grassmann variety and is an irreducible algebraic variety without singularities so that intersection can be well defined in it, see e.g.[HP], Chap.XI.

Following Ehresmann let us consider a fixed sequence of linear subspaces

$$S0-<S1-<\cdots-<Sn=CPn. \tag{1.1}$$

Let Ai, Bj be integers verifying

$$0<=A0<A1<\cdots<Ap<=n, \tag{1.2}$$

$$0<=B0<B1<\cdots<Bq<=n. \tag{1.3}$$

With respect to (1.1) we shall denote by the Ehresmann symbol of the form

$$[A0, A1, \cdots, Ap/B0, B1, \cdots, Bq] \tag{1.4}$$

the totality of pairs $([p], [q])$ such that

$$(E1) \quad \dim([p] \wedge [SAi]) >= i, \quad \text{for} \quad 0<=i<=p;$$

On Chern Numbers of Algebraic Varieties with Arbitrary Singularities

(E2) $\dim([q] \wedge [SBj]) >= j$, for $0 <= j <= q$;

(E3) each Ai is some Bj.

The variety (1.4), usually called a Schubert variety, has a dimension

$$\dim[A0, A1, \cdots, Ap/B0, B1, \cdots, Bq] = \text{SUM}i(Ai - i) + \text{SUM}'j\ (Bj - j). \tag{1.5}$$

In (1.5) SUMi is to be extended over i from 0 to p while SUM$'j$ is over only such j from 0 to q for which Bj is not equal to any Ai. In particular, $GR\ (n;\ p,\ q)$ is itself such a Schubert variety with symbol and dimension given by

$$GR(n; p, q) = [(n - p, \cdots, n)/(n - q, \cdots, n)], \tag{1.6}$$

$$\dim GR(n; p, q) = (n - p) * (p + 1) + (n - q) * (q - p). \tag{1.7}$$

We remark that the bracket ( ) in (1.6) means that the integers therein are consecutive ones.

Take now a second fixed sequence of linear subspaces

$$S'0- < S'1- < \cdots - < S'n = CPn \tag{1.1}'$$

for which all $S'i$ in (1.1)$'$ and Sj in (1.1) are in general position. Denote the Schubert variety corresponding to (1.4) defined however with respect to (1.1)$'$ by

$$[A0, A1, \cdots, Ap/B0, B1, \cdots, Bq]'. \tag{1.4}'$$

We shall set

$$[A'p, \cdots, A'1, A'0/B'q, \cdots, B'1, B'0]' = \text{Dual}[A0, A1, \cdots, Ap/B0, B1, \cdots, Bq],$$

in which $A'i = n - Ai$ and $B'j = n - Bj$. We see that any Schubert variety will intersect its dual in a single point.

According to Chow (see[CHOW]), the variety $GR(n;\ p,\ q)$ has a rational dissection formed of all the above Schubert varieties defined with respect to (1.1) with boundaries removed and the totality of such Schubert varieties will represent a basis of the group of rational equivalence classes of $GR(n;\ p,\ q)$. The rational dissection defined with respect to (1.1)$'$ is then said to be $DUAL$ to the rational dissection above defined with respect to (1.1) in the sense of [WU2]. It easily follows that the totality of Schubert varieties (1.4)(or (1.4)$'$) form also a basis of the group of algebraic equivalence classes of $GR(n;\ p,\ q)$. For an algebraic variety with arbitrary singularities V

let us denote by $RATr(V)$ respectively $ALGr(V)$ the group of rational respectively algebraic equivalence classes in dimension r of V. Denote also for any subvariety W of dimension r of V, its rational respectively algebraic equivalence class, by $R-Cls(W)$ respectively $A-Cls(W)$. If V is devoid of any singularity, then the sum of $ALGr(V)$ for all r will possess an intersection

$$ALGr(V) * ALGs(V) -< ALGt(V)$$

with $t = r + s - \dim V$ which turns the sum into an intersection ring or CHOW RING of the nonsingular variety V. In particular for the nonsingular $GR(n; p, q)$ we have for any Schubert varieties E, E' the formulae of intersection

$$A - Cls(E) * A - Cls \text{ Dual } E = 1, \text{ while } A - Cls(E) * A - Cls(E') = 0, \text{ for}$$
$$E' <> \text{Dual } E, \text{ and } \dim E' + \dim E = \dim GR(n; p, q).$$

Furthermore, the association of any Schubert variety to its dual will induce a natural morphism

$$\text{Dual}: ALGs(GR(n; p, q)) \to ALGt(GR(n; p, q)),$$

in which $t = \dim GR(n; p, q) - s$.

Sect 2. The intersection ring of GR (n; 0, d)

For the purpose of the present paper we shall restrict ourselves henceforth to the particular case of the composite grassmann $GR(n; p, q)$ with $p = 0, q = d$. The dimension of our grassmannian is then given by

$$\dim GR(n; 0, d) = (n - d) * (d + 1) + d.$$

For reasons to be explained later we are particularly interested in algebraic equivalence classes below:
$GAMst = A - Cls[s - t/(0, \cdots, d - t), (d - t + 2, \cdots, d + 1)] -< ALGs \ GR(n; 0, d)$
for $0 <= t <= s <= d$, and

$$CHs = \text{SUM}t(\text{sgn}(t) * (d - t + 1//d - s + 1) * GAMst) -< ALGs \ GR(n; 0, d)$$

for $0 <= s <= d$, in which SUMt means summation extended over t from 0 to s.

Remark In [WU1-3] there are some misprints in sign in the binomial coefficients.

For $s = 1$, 2 or 3 we have in particular

$$CH1 = (d+1) * GAM10 - GAM11$$
$$= (d+1) * A - Cls[1/(0, \cdots d)] - A - Cls[0/(0, \cdots, d-1), d+1], \qquad (2.1)$$

$$CH2 = d * (d+1)/2 * CAM20 - d * GAM21 + GAM22$$
$$= d * (d+1)/2 * A - Cls[2/(0, \cdots, d)] - d * A - Cls[1/(0, \cdots, d-1), d+1]$$
$$+ A - Cls[0/(0, \cdots, d-2), d, d+1]. \qquad (2.2)$$

$$CH3 = (d+1//3) * A - Cls[3/(0, \cdots, d)] - (d//2) * A - Cls[2/(0, \cdots, d-1), d+1]$$
$$+ (d-1) * A - Cls[1/(0, \cdots, d-2), d, d+1]$$
$$- A - Cls[0/(0, \cdots, d-3), d-1, d, d+1]. \qquad (2.3)$$

The intersection structure or Chow ring of $GR(n; 0, d)$ will only be partially determined but will be sufficient for our purposes. For this we shall first prove the following lemmas.

Lemma 1 *For the Schubert variety*

$$A = [n - Ai/n - Ad, \cdots, n - Ai, \cdots, n - A0] = \text{Dual } [Ai/A0, \cdots, Ai, \cdots, Ad]$$

to have a dimension $>= \dim GR(n; 0, d) - d$ *it is necessary that*

$$A0 = 0, A1 = 1, \cdots, Ai = i$$

so that

$$A = [n - i/n - Ad, \cdots, n - Aj, (n - i, \cdots, n)]$$

in which we have put $i + 1 = j$.

Proof. The hypothesis implies that

$$\dim[Ai/A0, \cdots, Ai, \cdots, Ad] = \text{SUM}k(Ak - k) + i <= d.$$

Now the integers Ak should verify the conditions

$$0 <= A0 < A1 < \cdots < Ad <= n, \text{ or}$$
$$0 <= A0 <= A1 - 1 <= \cdots <= Ai - i <= Aj - j <= \cdots <= Ad - d.$$

It follows that $Ai > i$ would imply $Ak > k$ for all $k >= i$ so that $\text{SUM}k\ (Ak-k)+i > d$ contractory to the inequality given above. Consequently $Ai = i$ and whence $A0 = 0, A1 = 1$, etc. up to $Ai = i$ as to be proved.

Lemma 2 Let $j = i+1$. Then

$$A - Cls[n-i/n-Ad, \cdots, n-Aj, (n-i, \cdots, n)]$$
$$= A - Cls[n/n-Ad, \cdots, n-Aj, (n-i, \cdots, n)] * A - Cls[n-i/(n-d, \cdots, n)]. \quad (2.4)$$

Proof. Denote the Schubert varieties involved in the above equality by A, B, C respectively defined with respect to sequences of linear subspaces like (1.1) as follows

For $B: [0]- < [1]- < \cdots - < [n] = CPn.$

For $C: [0]'- < [1]'- < \cdots - < [n]' = CPn$

with $[n-i-1]- < [n-i]'$ but otherwise the $[r]$ and $[s]'$ are in general position. The variety A is then defined with respect to the sequence
$[0]- < [1]- < \cdots - < [n-i-1]- < [n-i]'- < [n-i+1]'- < \cdots - < [n] = CPn$.
Clearly an element $(S0, Sd)$ of $GR(n; 0, d)$ will belong to the intersection of B and C if and only if

$$\dim(Sd \wedge [n-Ak]) >= d - k,$$

for $k = i+1, \cdots, d$ and $S0- < [n-i]'$, i.e.

$$(S0, Sd)- < A.$$

Now
$$\dim B = \dim GR(n; 0, d) - \text{SUM}k(Ak - k),$$

$\dim C = \dim GR(n; 0, d) - i$, and $\dim A = \dim GR(n; 0, d) - \text{SUM}k(Ak - k) - i$, in which SUM$k$ is to be extended over k from $i+1$ to d. Now by Ehresmann the variety A is an irreducible one. It follows then from dimensionality considerations that the right-hand side should be equal to an integral multiple of the class of A. This integer is the intersection multiplicity of B and C and is easily seen to be 1. This proves the formula (2.4) of the Lemma.

It is clear that

$$A - Cls[n-i/(n-d, \cdots, n)] * A - Cls[n-j/(n-d, \cdots, n)]$$
$$= A - Cls[n-i-j/(n-d, \cdots, n)]. \quad (2.5)$$

Furthermore we have also in the right dimensions

$$A - Cls[n/n-Ad, \cdots, n-A0] * A - Cls[n/n-Bd, \cdots, n-B0]$$

$$= \text{SUMc } A - Cls[n/n - Cd, \cdots, n - C0], \tag{2.6}$$

in which

$$A - Cls[n - Ad, \cdots, n - A0] * A - Cls[n - Bd, \cdots, n - B0]$$
$$= \text{SUMc } A - Cls[n - Cd, \cdots, n - C0] \tag{2.7}$$

is the intersection formula in the ordinary grassmannian as shown in [HP], Chap. XIV, which can in turn be explicitely determined by means of the well-known formulae of Pieri and Giambelli.

Let us now introduce some classes as follows:

$$P = A - Cls[1/(0, \cdots, d)], \tag{2.8}$$
$$Qh = A - Cls[0/(0, \cdots, d-1), d+h], \tag{2.9}$$
$$P' = \text{Dual } P = A - cls[n - 1/(n - d, \cdots, n)], \tag{2.8}'$$
$$Q'h = \text{Dual } Qh = A - Cls[n/n - d - h, (n - d + 1, \cdots, n)] \tag{2.9}'$$

for $0 <= h <= n-d$. From the above we get easily the following.

Theorem *The Chow ring of algebraic equivalence classes of the composite grassmannian $GR(n; 0, d)$ is generated by the classes P' and $Q'h$ in the dimension $>= \dim GR(n; O, d) - d$. The multiplicative structure in that range is completely determined by the formulae (2.4) – (2.9)'.* By the theorem we deduce from (2.1) – (2.3):

$$\text{Dual } CH1 = (d+1) * P' - Q'1, \tag{2.10}$$
$$\text{Dual } CH2 = d * (d+1)/2 * P' \wedge 2 - d * P' * Q'1 + (Q'1 \wedge 2 - Q'2), \tag{2.11}$$
$$\text{Dual } CH3 = (d+1) * d * (d-1)/6 * P' \wedge 3 - d * (d-1)/2 * P' \wedge 2 * Q'1$$
$$+ (d-1) * P' * (Q'1 \wedge 2 - Q'2) - (Q'1 \wedge 3 - 2 * Q'1 * Q'2 + Q'3). \tag{2.12}$$

Sect 3. Ehresmann classes of an algebraic variety with arbitrary singularities

Let Vd be an irreducible algebraic variety of dimension d and V' a subvariety containing all singularities of Vd. By considering subvarieties of a fixed dimension s, the author has introduced in [WU1] the notion of group of UNNEGLIGIBLE algebraic equivalence classes modulo V' for each dimension s, with methods as described in Chap. XI of [HP], which will be denoted by $ALGs(Vd/V')$ in what follows. There is also a natural morphism for each dimension s, viz.

$$Js : ALGs(Vd/V') \to ALGs(V).$$

Let Wd be also some irreducible algebraic variety of same dimension d with W' a subvariety containing all singularities of Wd. Let T be a birational transformation of Wd to Vd verifying the following properties:

P1. T is everywhere defined on Wd.

P2. $T(x)- < V'$ if and only if $x- < W'$.

P3. T is biunivoque on $Wd - W'$.

It is proved in[WU1]that under these conditions the birational transformation T will induce in each dimension s a natural morphism

$$Ts : ALGs(Wd/W') \to ALGs(Vd/V').$$

Note that for these groups of unnegligible algebraic equivalence classes no mulplicative structure will be introduced in their sum.

Let Ge be now an irreducible algebraic variety of dimension e in a complex projective space with no singularities so that intersection may be defined in Ge in the usual manner. Let Wd be an irreducible subvariety of dimension d in Ge and W' a subvariety of Wd containing all singularities of Wd if exist. As Ge is in a complex projective space any subvariety of Ge is algebraically equivalent to some one which will intersect simply with both Wd and W'. From this we easily deduce that, by considering intersections with Wd in Ge, there will be natural morphisms

$$Is : ALGs(Ge) \to ALCt(Wd/W'),$$

in which $t = s + d - e$.

Consider now an irreducible algebraic variety Vd of dimension d in a projective space CPn of dimension n. Take an arbitrary generic point $P0$ of Vd and let Pd be the tangent space of Vd at $P0$. With the pair $(P0, Pd)$ as a generic point there will be a determined irreducible subvariety Wd of dimension d in the composite grassmannian $GR\,(n; 0, d)$ which may he considered as a subvariety of a projective space of sufficiently high dimension. Now any pair $(P0', Pd')$ of Wd is a specialization of

$(P0, Pd)$ which implies that $P0'$ is a specialization of $P0$ and is thus a well-defined point of Vd. Clearly, if the singular subvariety of Vd is V', then the subvariety W' of Wd consisting of all points $(P0', Pd')$ with $P0'$ in V' will contain all the singular points of Wd if there are any. The correspondence

$$T : (P0', Pd') \longrightarrow P0'$$

is thus a birational one verifying the properties P1-3 with $V' = T(W')$. For the pair $G = GR(n; 0, d)$ and Vd we have then a sequence of morphisms

$$ALGs(GR(n;O,d)) \xrightarrow{Is} ALGt(Wd/W') \xrightarrow{T} ALGt(Vd/V') \xrightarrow{Jt} ALGt(Vd),$$

in which $t = s + d - e$, with $e = \dim GR(n; 0, d)$. Besides we have also the dual morphism

$$\text{Dual} : ALGs(GR(n; 0, d)) \to ALGs'(GR(n; 0, d))$$

in which

$$s' = \dim GR(n; 0, d)) - s.$$

Consider now any Ehresmann symbol

$$EH = [A0/B0, B1, \cdots, Bd] \quad \text{with} \quad s = SUM'k(Bk - k) + A0$$

and

$$r = s' + d - \dim GR(n; 0, d) = d - s,$$

in which $SUM'k$ is to be extended over k from 0 to d for which $Bk <> A0$. We shall lay down the following

Definition The algebraic equivalence class

$$Jr \ T \ Is' \ \text{Dual} \ EH - < ALGr(Vd)$$

will be called the EHRESMANN CLASS of Vd corresponding to the symbol EH and will be denoted by

$$EH(Vd) = [A0/B0, B1, \cdots, Bd](Vd).$$

More generally, for any algebraic equivalence class $ACLS - < ALGs(GR(n; O, d))$, we shall set by definition

$$ACLS(Vd) = Jr \ T \ Is' \ \text{Dual} \ ACLS - < ALGr(Vd).$$

As particular Ehresmann classes we have also GAMKRELIDZE CLASSES and CHERN CLASSES defined respectively by $(r = d - s)$

$GAMst(Vd) = [s - t/(0, \cdots, d - t), (d - t + 2, \cdots, d + 1)](Vd)- < ALGr(Vd)$,
$CHs(Vd) = \text{SUM}t(\text{sgn}(t) * (d - t + 1)//(d - s + 1) * GAMst(Vd))- < ALGr(Vd)$.

We note that in case that Vd is devoid of any singularities so that Vd may be considered as a complex manifold in a complex projective space, then according to Gamkrelidze the homology classes defined by the algebraic equivalence classes $CHs(Vd)$ are just the dual of the usual Chern classes. This justifies the terminologies introduced above, cf. [G].

Remark that in the notations $EH(Vd), CHs(Vd)$, etc., integer n, the dimension of the ambiant projective space in which lies the variety Vd, does not enter into play, as is natural and easy to see.

Sect 4. Chern numbers of an algebraic variety with arbitrary singularities

Let Vd be an irreducible algebraic variety of dimension d with arbitrary singularities. Consider any Ehresmann symbol

$$EH = [A0/B0, B1, \cdots, Bd] \text{ with } \text{SUM}'k(Bk - k) + A0 = d,$$

in which $\text{SUM}'k$ is to be extended over k from 0 to d for which $Bk <> A0$. The Ehresmann class $EH(Vd)$ of $ALG0(Vd)$ is in the image of $ALG0(Vd/V')$ under the morphism $J0$ and can thus be identified to an integer, to be called the EHRESMANN CHARACTER of Vd corresponding to the symbol EH in what follows. From the definition it is clear that all such characters are of projective nature and were known as PROJECTIVE CHARACTERs of the variety in the sense of Severi, cf.e.g. [SR]. Among these Ehresmann characters we have in particular CHERN CHARACTERs to be defined as follows.

A sequence of integers $p = (a, b, \cdots, c)$ will be said to be a partition of d if

$$0 < a <= b <= \cdots <= c, \text{ and } a + b + \cdots + c = d.$$

Define now $CHp- < ALGd(GR(n; 0, d))$ by

$$\text{Dual } CHp = \text{Dual } CHa * \text{Cual } CHb * \cdots * \text{Dual } CHc.$$

The integer identified to the algebraic equivalence class $CHp(Vd) - < ALG0(Vd)$ will then be called the CHERN CHARACTES of Vd corresponding to the partition p. By the intersection formulae in $GR(n; 0, d)$ as developed in the preceding sections it is clear that any such CHp can be expressed by means of algebraic equivalence classes P and Qh.

Let us consider as an example the case $d = 2$, i.e. the case of an algebraic surface $V2$ with arbitrary singularities. For such a $V2$ we have 4 Ehresmann characters and 2 Chern classes besides the trivial one $CH0$, viz.

$$[2/0, 1, 2](V2) = \text{Classical } Mu0(V2) = \text{Order of } V2,$$
$$[1/0, 1, 3](V2) = \text{Classical } Mu1(V2) = \text{Rank of } V2,$$
$$[0/0, 2, 3](V2) = \text{Classical } Mu2(V2) = \text{Class of } V2,$$
$$[0/0, 1, 4](V2) = \text{Classical } Nu2(V2) = \text{Type of } V2.$$

The last terminology is for $V2$ in CPn with $n > 3$ alone, but we shall keep this term for $V2$ in $CP3$ too. Cf. [SR], Chap. IX.

$$CH1(V2) = 3 * [1/0, 1, 2](V2) - [0/0, 1, 3](V2) = (\text{Dual}(3*P' - Q'1))(V2),$$
$$CH2(V2) = 3 * [2/0, 1, 2](V2) - 2 * [1/0, 1, 3](V2) + [0/0, 2, 3](V2)$$
$$= (\text{Dual}(3 * P' \wedge 2 - 2 * P' * Q'1 + Q'1 \wedge 2 - Q'2))(V2).$$

There are 2 Chern characters $CH11$ $(V2)$ and $CH2$ $(V2)$ for which we have for the former

$$CH11(V2) = (\text{Dual}(9 * P' \wedge 2 - 6 * P' * Q'1 + Q'1 \wedge 2))(V2).$$

It follows that

$$3 * CH2(V2) - CH11(V2) = 2 * (\text{Dual } Q'1 \wedge 2)(V2) - 3 * (\text{Dual } Q'2)(V2).$$

Now in the grassmannian $GR(n; 0, d)$ we have the multiplication formula

$$A - Cls[n/n - 3, n - 2, n] = A - Cls[n/n - 3, n - 1, n]$$
$$\wedge 2 - A - Cls[n/n - 4, n - 1, n].$$

Taking the dual of both sides we get

$$[0/0, 2, 3](V2) = (\text{Dual } Q'1 \wedge 2)(V2) - (\text{Dual } Q'2)(V2).$$

As the left side is $Mu2(V2)$ and the last term is $Nu2(V2)$ we get

$$2 * (\text{Dual } Q'1 \wedge 2)(V2) - 3 * (\text{Dual } Q'2)(V2) = 2 * Mu2(V2) - Nu2(V2).$$

If $V2$ is in $CP3$ then $Nu2(V2)$ is clearly 0 and hence we get the following

Theorem For a surface $V2$ in $CP3$ with arbitrary singularities we have for the Chern characters the inequality

$$3 * CH2(V2) >= CH11(V2).$$

Suppose that the surface $V2$ has no singularities so that it is a SMOOTH complex surface. Then $CH2(V2)$ is just the usual Chern number $c_2(V2)$ and $CH11(V2)$ the Chern number $c_1^2(V2)$. The above inequality becomes then the Miyaoka-Yau inequality stated in the beginning of the paper. The above theorem can therefore be considered as a generalization of the Miyaoka-Yau inequality to the case of algebraic surfaces with arbitrary singularities lying in $CP3$.

On the other hand suppose that the variety $V2$ is not in $CP3$. Then there are known examples for which

$$2 * Mu2(V2) < Nu2(V2).$$

Cf. formulae (10) and (11) on [SR], p.221. It follows that the Miyaoka-Yau inequality is not true in general for surfaces in CPn with singularities present. We leave open the question of the truth of the inequality in case of NON-SINGULAR $V2$ in CPn with $n > 3$.

Consider now any hypersurface Vd of dimension d in CPn with $n = d + 1$. We have then from the very definition

$$Q'h = 0 \text{ for } h >= 2.$$

For $d = 3$ in particular we would have then from (2.10)—(2.12):

Dual $CH1 = 4 * P' - Q'1$,
Dual $CH2 = 6 * P' \wedge 2 - 3 * P' * Q'1 + Q'1 \wedge 2$,
Dual $CH3 = 4 * P' \wedge 3 - 3 * P' \wedge 2 * Q'1 + 2 * P' * Q'1 \wedge 2 - Q'1 \wedge 3$.

There are 3 partitions $(1, 1, 1)$, $(1, 2)$ and (3) of the integer $d = 3$ for which we have

Dual $CH111 = (4 * P' - Q'1) \wedge 3 = 64 * P' \wedge 3 - 48 * P'$
$\wedge 2 * Q'1 + 12 * P' * Q'1 \wedge 2 - Q'1 \wedge 3$,
Dual $CH12 = (4 * P' - Q'1) * (6 * P' \wedge 2 - 3 * P' * Q'1 + Q'1 \wedge 2)$
$= 24 * P' \wedge 3 - 18 * P' \wedge 2 * Q'1 + 7 * P' * Q'1 \wedge 2 - Q'1 \wedge 3$.

whence

$$4 * CH12(V3) - 8 * CH3(V3) - CH111(V3) = 5 * (\text{Dual } Q'1 \wedge 3)(V3).$$

As (Dual $Q'1 \wedge 3$)($V3$) is necessarily non-negative we get the following generalization of a theorem due to Tai (cf. [T]), viz.

Theorem   For a hypersurface V3 of dimension 3 in CP4 with arbitrary singularities we have for the Chern characters the inequality

$$4 * CH12(V3) - 8 * CH3(V3) - CH111(V3) >= 0.$$

Clearly the method is entirely general which will permit us to get generalizations of other theorems of Tai to case of algebraic hypersurfaces with arbitrary singularities. We can also investigate possible generalizations of inequalities of Miyaoka-Yau type in the case of higher dimensions. We shall however not enter into these problems since the method of treatment is quite clear.

References

[CH] Chern, S.S. On the characteristic classes of complex sphere bundles and algebraic varieties. *Amer. Math.Soc.*, 1953, 75: 565-597.

[CHOW] Chow W.L. Algebraic varieties with rational dissections. *Proc. Nat. Acad. Sci.*, 1956, 42: 116-119.

[EH] Ehresmann, C. Sur la topologie de certaines espaces homogenes. *Ann. of Math.*, 1934, 35: 396-443.

[G] Gamkrelidze, P.B. Chern cycles of complex algebraic manifolds. *IZV. ACAD. SCIS, CCCP, Math. Ser.*, 1956, 20: 685-706 (in Russian).

[HP] Hodge, W. V. D. and Pedoe, D. *Methods of Algebraic Geometry*. Cambridge, 1947, 1; 1952, 2.

[MP] MacPherson, R.D. Chern classes for singular algebraic varieties. *Ann. of Math.*, 1974, 100: 423-432.

[M] Miyaoka. Y. On the Chern numbers of surfaces of general type. *Invent. Math.*, 1977, 42: 225-237.

[SR] Semple, J.G. and Roth, L. *Introduction to Algebraic Geometry*. Oxford, 1949.

[T] Tai, S. A class of symmetric functions and Chern numbers of algebraic varieties. Preprint, 1985.

[VdV] Van de Ven. On the Chern numbers of surfaces of general type. *Invent. Math.*, 1976, 36: 285-293.

[VdW] Van der Waerden. *Einfuerung in die Algebraischen Geometric*. Berlin, 1945.

[WU1] Wu Wen-tsun. On Chern characteristic systems of an algebraic variety (in Chinese). *Shuxue Jinzhan*, 1965, 8: 395-401.

[WU2] —. On algebraic varieties with dual rational dissections (in Chinese). *Shuxue Jinzhan*, 1965, 8: 402-409.

[WU3] —. Chern classes on algebraic varieties with arbitrary singularities, in *Several Complex Variables*. Proc. 1981 Hangzhou Conf., Eds. Kohn et al. Birkhauser, 1984: 247-249.

[Y] Yau, S.T. Calabi's conjecture and some new results in algebraic geometry. *Proc. Nat. Acad. Sci.*, 1977, 74: 1798-1799.

集成电路设计中的一个数学问题[*]

§1. 引言

集成电路牵涉到线性图的平面性即是否可在平面中拓扑地实现的问题,这个问题的各个方面可用下图来概括:

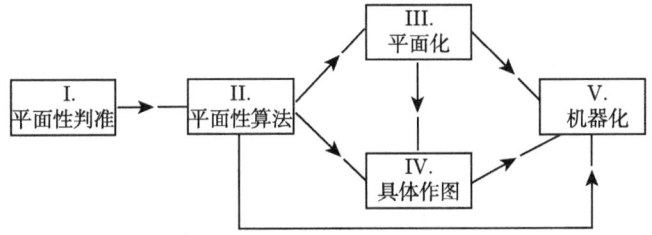

I—V 的几个方面可说明如下:

I.**平面性判准**—— 给出一个线性图是否有平面性的判准. 这样的判准早在 30 年代就已给出,如 Kuratowski[4], Whitney[6], MacLane[5] 等. 其中 Kuratowski 判准说,线性图 G 有平面性的充要条件是, G 不含有任何下面两类子图形之一:

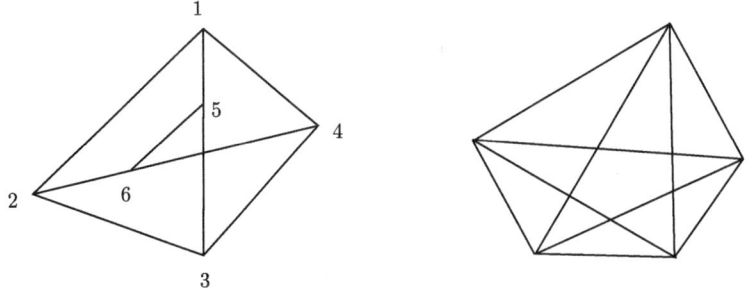

但这一类判准在实用上有困难,例如下线性图是非平面性的,有一 Kuratowski 子图形如粗黑线所示,但并不容易找出:

---

[*] 本文原载《数学的实践与认识》,1973 年第 6 期,20-40. 本文在工作过程中,曾得到冯康与吴稽康二同志不少帮助,又承数学所虞言林同志仔细校验一遍,并指出谬误数处,谨此致谢.

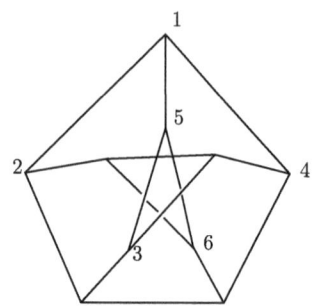

Ⅱ．**平面性算法**—— 给出一个算法，依据它能依照确定步骤来判断一个线性图是否有平面性．上面所提到的那些判准都不足以从此导出一个算法来．这样的算法直到近几年估计是工程技术问题上的要求才出现，见 Auslander-Parter[1]，与 Bader[2]．

Ⅲ．**平面化**—— 原来的线性图是非平面性的，但在除去某些棱后余下的线性图可以是平面性的．这种手续可叫做线性图的*平面化*．

Ⅳ．**具体作图**—— 即使线性图 G 有平面性或已平面化，也并不等于说 G 已在平面上画出．因之需要有一个将 G 在平面上具体画出的方法．

Ⅴ．**机器化**—— 由于所考虑到的线性图往往是很复杂的，即使Ⅱ—Ⅳ诸问题在原则上已解决也难以用手算来付诸实施．因之需要程序化，使Ⅱ—Ⅳ各问题都能在电子计算机上解决．

国外对这些问题的处理情况如下：

Auslander-Parter[1] 与 Bader[2]（两文内容实质上是相同的）给出了一个分解原来线性图为更小的图形并重复进行的方法来解决问题Ⅱ—Ⅳ，Fisher-Wing[3] 更依据它来机器化．但是他们所给出的只是一个具体进行的方法而缺少理论的提炼，而 Kuratowski 等则虽有理论却不能实际使用．

我们对这些问题的处理情况如下：

对问题Ⅰ，我们在 1954 年以来所创立的示嵌类理论早已提供了一个简单的判准．这个判准容易化为一组模 2 系数的线性方程组的可解性问题，在这组方程有解时，根据它的解即可具体作图．这给出了问题Ⅱ—Ⅳ的解答，依据它来机器化乃是一个纯粹程序设计的问题，并没有什么实质性的困难．

§2. 线性图的平面性与示嵌类

设 G 是一线性图，G 的顶点的集合将记作 G^0，棱的集合记作 G^1．我们将作下述假定，这对 G 的平面性的考虑不产生影响：

$1°$．没有两端是同一顶点的棱．

$2°$. 连接 G 中任两顶点的棱, 至多只有一条.

$3°$. G 是连通的.

记 G^1 中的棱为 $e_1, \cdots, e_n$, 指数集 $\{1, \cdots, n\}$ 记作 I. 对于一切无序指数偶 $(i,j), i \neq j, i, j \in I$, 我们将分成两部分 D 与 N. 如果 e_i, e_j 在 G 中无公共顶点, 则置 $(i,j) \in D$, 否则置 $(i,j) \in N$.

对称约化积

对 G 可作一二维复形 G^*, 称之为 G 的二重对称约化积者如次, G^* 的二维胞腔对应于一切 $(i,j) \in D$, 记作 $e_i * e_j = e_j * e_i$. G^* 的一维胞腔对应于 G 中一切顶点 v 与棱 e_i 的偶合, 而 v 非 e_i 的顶点, 记之为 $v * e_i = e_i * v$. G^* 的 0 维胞腔对应于 G 中一切无序的顶点偶 (v,w), 而 $v \neq w$, 记之为 $v * w = w * v$. 如果 e_i 的顶点是 v_{i1}, v_{i2}, e_j 的顶点是 v_{j1}, v_{j2}, 则 G^* 中的 (模 2) 边界关系将定义为

$$\begin{cases} \partial(e_i * e_j) = v_{i1} * e_j + v_{i2} * e_j + v_{j1} * e_i + v_{j2} * e_i, \\ \partial(v * e_i) = v * v_{i1} + v * v_{i2}, \\ \partial(v * w) = 0. \end{cases}$$

如果引入以下规定:

$$\begin{cases} e_i * e_j = 0, \ (i,j) \in N \text{ 时}; v * e_i = 0, v \text{ 为 } e_i \text{ 端点时}; v * v = 0, \\ \sum_{i,j} \alpha_{ij} e_i * e_j = \sum_j \left(\sum_i \alpha_{ij} e_i \right) * e_j = \sum_i \left(e_i * \sum_j \alpha_{ij} e_j \right), \end{cases}$$

余类推. 则以上边界关系亦可表作 $(e_i * e_j, v * e_i, v * w \in G^*)$

$$\begin{cases} \partial(e_i * e_j) = \partial e_i * e_j + e_i * \partial e_j, \\ \partial(v * e_i) = v * \partial e_i, \\ \partial(v * w) = 0. \end{cases}$$

或对偶地说,

$$\begin{cases} \delta(v * w) = \delta v * w + v * \delta w, \\ \delta(v * e_i) = \delta v * e_i, \\ \delta(e_i * e_j) = 0. \end{cases}$$

浸入与嵌入

以 R^2 表平面. 映象 $f : G \to R^2$ 将称为一个**浸入**, 记为 $f : G \subset R^2$, 如果以下条件满足:

(a) 对每一棱 $e_i \in G^1$, f/e_i 是一拓扑映象.
(b) 对每一棱 $e_i \in G^1$, $f(e_i)$ 除原来的端点外, 不再经过其他顶点的象.
(c) 对每两棱 $e_i \neq e_j \in G^1$, $f(e_i), f(e_j)$ 除可能有的公共顶点外, 至多"简单相交"于有限个内点, 即每一交点处的相交指数为 1(模 2).

如果浸入 f 对每两棱 $e_i \neq e_j \in G^1$, $f(e_i), f(e_j)$ 至多在可能有的公共顶点处相遇, 则 f 是 G 的一个拓扑映象. 此时 f 将称为一个嵌入, 记作 $f: G \subset R^2$. 如果有这样的嵌入存在, 则 G 称为一平面图形, 或具有平面性, 记作 $G \subset R^2$.

(注: 这里的浸入定义较 [7] 中所引入者为弱, 原来尚须要求满足下述条件.)

(d) 对每二有公共顶点 v 的棱 $e_i, e_j \in G^1$, $f(e_i), f(e_j)$ 只有一公共点 $f(v)$.

示嵌链

所有的链, 同调等都将是模 2 的, 在以下不再说明.

设 $f: G \subset R^2$ 是一个浸入. 对 f 引入一 G^* 中的二维上闭链 $\varphi_f = \varphi$ 如次. 对任意 $(i,j) \in D$, 命
$$\varphi_{ij} = I_f(e_i, e_j),$$
即 $f(e_i), f(e_j)$ 的相交指数. 置
$$\varphi = \sum_{(i,j) \in D} \varphi_{ij} e_i * e_j.$$

显然 φ 是上闭链. 我们将称 φ 为浸入 f 的**示嵌链**. φ 之不等于 0 可视为 f 异于嵌入的一个测度.

定理 对任意浸入 f, 相应的示嵌链 φ 恒属于同一个上同调类 Φ.

证. 设 $f, g: G \subset R^2$, 相应的示嵌链各为 $\varphi = \sum \varphi_{ij} e_i * e_j$ 与 $\psi = \sum \psi_{ij} e_i * e_j$. 需证 $\varphi \sim \psi$.

为此, 先设对任意顶点 $v, f(v) = g(v)$, 且有一 $e_k \in G^1$, 在 $e_i \neq e_k$ 时, $f \equiv g/e_i$, 而 $f(e_k), g(e_k)$ 除公共端点外不相遇. 于是 $f(e_k), g(e_k)$ 构成一 R^2 中的简单闭曲线 C. 设在 f 与 g 下象在 C 内部的顶点为 $v_{k1}, \cdots, v_{ks}$. 试证

$$(*) \qquad \varphi = \psi + \sum_{r=1}^{s} \delta v_{kr} * e_k.$$

为此, 先设 $i \neq k, j \neq k$, 而 $(i,j) \in D$, 则显有 $\varphi_{ij} = \psi_{ij}$ 而 $(*)$ 式左右两边 $e_i * e_j$ 的系数相等. 今设 $(k,l) \in D$, 而 e_l 两端点在 f, g 下的象同在 C 的内部. 则此两端点将为一 v_{kp} 与一 v_{kq}. 于是在 $(*)$ 式右边的 $\sum$ 中 $e_k * e_l$ 将出现两次而其和为 0. 其次 $f(e_l) = g(e_l)$ 与 C 将简单相交于偶数次, 因而 $\varphi_{kl} = \psi_{kl}$. 故此时两边 $e_k * e_l$ 的系数仍相等. 当两端点在 f, g 下的象同在 C 外部时亦如此. 最后设两端点的象

一在 C 内而一在 C 外, 在内者设为 v_{kp}. 则 (*) 式右边 $\sum$ 中 $e_k * e_l$ 的项将出现一次. 又此时 $f(e_l) = g(e_l)$ 将与 C 简单相交于奇数个点, 因而 $\varphi_{kl} = \psi_{kl} + 1$. 故两边 $e_k * e_l$ 的系数仍将相等. 因而 (*) 式成立而有 $\varphi \sim \psi$.

在上情形中 $f(e_k), g(e_k)$ 除公共顶点外不相遇的条件容易除去. 盖可取 $h : G \subset R^2$ 使 $e_i \neq e_k$ 时, $h \equiv f \equiv g/e_i$, 而 $h(e_k), f(e_k)$ 以及 $h(e_k), g(e_k)$ 各满足上述条件. 命 h 的示嵌链为 θ, 则将有 $\theta \sim \varphi$ 与 $\theta \sim \psi$, 因而 $\varphi \sim \psi$.

次设对任意顶点 v 有 $f(v) = g(v)$ 而对任意两棱 $e_k, e_l \in G^1 (k$ 亦可 $= l), f(e_k)$ 与 $g(e_l)$ 除可能的公共端点外, 只简单相交于有限个内点, 即在每点的相交指数为 1(模 2). 今作 $h_0, \cdots, h_{n-1} : G \subset R^2$ 如次. $h_k : G \subset R^2$ 使 h_k 在 $e_1, \cdots, e_k$ 上与 f 相同而在 $e_{k+1}, \cdots, e_n$ 上与 g 相同. 则 $h_0 = g$ 而 $h_n = f$. 记 h_k 的相应示嵌链为 φ_k. 由已证明的情形, 应有 $\varphi_k \sim \varphi_{k+1}, k = 0, 1, \cdots, n-1$. 故得 $\varphi_0 \sim \varphi_n$ 即 $\psi \sim \varphi$.

若对任意顶点 v 仍有 $f(v) = g(v)$. 则总可作一浸入 $h : G \subset R^2$ 使对任意顶点 v, 有 $h(v) = f(v) = g(v)$, 而对任意两棱 $e_k, e_l \in G^1 (k$ 亦可 $= l), f(e_k)$ 与 $h(e_l)$ 以及 $g(e_k)$ 与 $h(e_l)$ 除可能的公共端点外, 只简单相交于有限个内点. 记 h 的相应示嵌链为 θ, 则应有 $\varphi \sim \theta, \psi \sim \theta$, 故仍有 $\varphi \sim \psi$.

最后, 设 f, g 任意. 命 h 为 R^2 到 R^2 上的一个拓扑变换, 使对任意顶点 $v \in G^0$, 有 $hf(v) = g(v)$. 记 hf 的相应示嵌链为 θ, 则依前述情形有 $\theta \sim \psi$, 另一面因 h 为拓扑变换故显然有 $\theta \sim \varphi$. 因之仍有 $\varphi \sim \psi$.

至此定理证毕.

定义 定理中所确定的上同调类 Φ 将称为 G 的 **示嵌类**.

如果 G 有平面性而 $f : G \subset R^2$ 是一嵌入, 则相应于 f 的示嵌链显为 0. 因而可得下述

嵌入定理 (必要部分) $\quad$ G 有平面性的一个必要条件是: G 的示嵌类 $\Phi = 0$, 或对任意浸入 $f : G \subset R^2$, 相应的示嵌链 $\varphi \sim 0$.

(注: 这个定理中的条件不仅是必要的, 而且是充分的. 但这一点将在 §5 中证明.)

§3. G 的平面性判准——基本定理与基本方程组

为了根据 §2 嵌入定理以确定 G 是否具有平面性, 试记 G 中顶点的集合为 $\{v_1, \cdots, v_m\} = G^0$. 任作一浸入 $f : G \subset R^2$, 记其相应的示嵌链为

$$\varphi = \sum \varphi_{ij} e_i * e_j.$$

于是 $G \subset R^2$ 的一个必要条件是 $\varphi \sim 0$ 或有 x_{λ_i} 存在使

$$\varphi = \sum x_{\lambda_i} \delta(v_\lambda * e_i),$$

这里的 $\sum$ 展开于一切使 v_λ 非 e_i 端点的指数偶 (λ,i) 上, 比较两边 e_i*e_j 的系数, 即得一组以 $x_{\lambda i}$ 为未知数的模 2 线性方程组. 这组方程可解是 G 有平面性的必要条件, 以后也将证明这些条件是充分的.

这个判准的原理虽然简单, 但由于未知数的个数接近于 mn, 方程的个数接近于 n^2, 因之当图 G 比较复杂, m, n 相当大时, 这个判准事实上并不实用, 即使用了电子计算机也是如此. 因之有必要简化上述方程组使之即使在 m, n 很大时也可以通过电子计算机来实现. 这是以下的目的所在.

最大树

为了这一目的, 试在 G 中取一树 T, 通过 G 的所有顶点. 这样的树 T 将称为 G 的一个最大树. T 的任一棱将称为一个节段, 顶点也称节点. 如果 T 中在节点 v_λ 处只有一个节段, 则 v_λ 也称为节梢.

对 T 的任一节段 e_i, T-内部 (e_i) 分成两个分支, 记为 C_i', C_i''. 命 $\sum_i$ 为所有 $e_a \in G^1 - T^1$ 的集合, 这里 e_a 的一端在 C_i' 中, 而另一端在 C_i'' 中. 又对任一棱 $e_a \in G^1 - T^1$, 在 T 中将有唯一的一条通路从 e_a 的一端到 e_a 的另一端, 记之为 $\mathscr{P}_a$. 置

$$E_i = \sum_{e_a \in \sum_i} e_a, \quad (e_i \in T^1),$$

$$P_a = \sum_{e_i \in \mathscr{P}_a} e_i, \quad (e_a \in G^1 - T^1),$$

这里 E_i, P_a 都是 G 中的 (模 2) 上链. 于是显然有

$$e_a \in \sum\nolimits_i \iff e_i \in \mathscr{P}_a.$$

引理 1 对于最大树 T 而言, G 的一维上边缘群有一组基底由以下各上边缘所构成:

$$e_i + E_i, \quad e_i \in T.$$

证. 对 $e_i \in T$, 命 $A_i'(A_i'')$ 为 $C_i'(C_i'')$ 中一切顶点的集合, 则易见

(1) $$e_i + E_i = \sum_{v' \in A_i'} \delta v' = \sum_{v'' \in A_i''} \delta v''.$$

故每一 $e_i + E_i$ 为一上边缘, 且不同的 $e_i + E_i$ 恰含有一不同的 $e_i \in T$, 故这些 $e_i + E_i$ 是线性独立的.

另一面, 对任一 G 的顶点 v, 设 $e_j, e_{j1}, \cdots, e_{j2}$ 是 T 中在 v 处的节段全体. 命 C_j' 为 T-内部 (e_j) 中含有 v 的分支, 而 C_{js}' 为 T-内部 (e_{js}) 中不含 v 的那一

分支. 则由 (1) 式有

$$\delta v = (e_j + E_j) + \sum_{s=1} (e_{js} + E_{js}).$$

故任一上边缘可表为形如 $e_i + E_i$ 的上边缘之和. 由此引理得证.

(注: 上述引理事实上是已知的, 在电网络理论中 $e_i + E_i$ 是一割集 (cut set), 参阅例如: S.Seshu-N.Balanian, *Linear Network Analysis*, 1959, 页 70—71.)

引理 2 G^* 的二维上边缘群有以下一组生成元:

$$\begin{cases} (e_i + E_i) * (e_j + E_j), & e_i, e_j \in T^1, (i,j) \in D \cup N; \\ (e_i + E_i) * e_\alpha, & e_i \in T^1, e_\alpha \in G^1 - T^1, (i,\alpha) \in D \cup N. \end{cases}$$

证. 命 v 为 G 的任一顶点, e 为 G 的任一棱, 而 v 非 e 的顶点, 则 G^* 的二维上边缘群由诸 $\delta(v*e) = \delta v * e$ 所生成. 因 G 的一维上链群有一组基由诸 $e_j \in T^1$ 与 $e_\alpha \in G^1 - T^1$ 所构成, 故也有一组基由诸 $e_j + E_j$ 与 e_α 所构成. 另一面由引理 1, δv 又可表为诸 $e_i + E_i, i \in T^1$ 的线性和, 故得本引理.

定义 对最大树 T 而言, 一个浸入 $f: G \subset R^2$ 将称为一个 T-浸入, 如果 (i) $f: T \subset R^2$, (ii) 对任意 $e_i \in T^1$ 与 $e_\alpha \in G^1 - T^1, f(e_i)$ 与 $f(e_\alpha)$ 至多在公共端点处相遇.

引理 3 设 $\tilde{T}$ 是 T 在 G 中的一个邻域, 则一个 T-浸入 f 的示嵌链由 $f/\tilde{T}$ 所完全决定. 换言之, 若 f, g 是两个 T-浸入, 而 $f/\tilde{T} \equiv g/\tilde{T}$, 则 $\varphi_f = \varphi_g$.

证. 若 $e_i, e_j \in T^1, e_\alpha \in G^1 - T^1$, 而 $(i,j), (i,\alpha) \in D$, 则 $\varphi_f(e_i * e_j) = \varphi_g(e_i * e_j) = 0, \varphi_f(e_i * e_\alpha) = \varphi_g(e_i * e_\alpha) = 0$. 今设 $e_\alpha, e_\beta \in G^1 - T^1, (\alpha, \beta) \in D$. 于是 $f(e_\alpha), f(\mathscr{P}_\alpha)$ 构成一简单闭曲线 C. 记 e_β 的两端为 α_1, α_2, 则 $f(\alpha_1) = g(\alpha_1), f(\alpha_2) = g(\alpha_2)$. 在 e_β 内部取两点 $\alpha_1', \alpha_2' \in \tilde{T}$ 充分接近于 α_1, α_2. 则视 α_1', α_2' 同在 C 之内或外或一在 C 内一在 C 外, 相交系数 $I(f(e_\beta), f(e_\alpha))$ 与 $I(g(e_\beta), f(e_\alpha))$ 将同为 0 或同为 1. 同样, 相交系数 $I(g(e_\beta), f(e_\alpha))$ 与 $I(g(e_\beta), g(e_\alpha))$ 亦同为 0 或 1. 故

$$I(f(e_\beta), f(e_\alpha)) = I(g(e_\beta), g(e_\alpha))$$

或

$$\varphi_j(e_\alpha * e_\beta) = \varphi_g(e_\alpha * e_\beta).$$

因而 $\varphi_f = \varphi_g$, 如所欲证.

基本定理 I (必要部分) G 有平面性的一个必要条件是, 对最大树 T 而言, 下述方程组有解:

(I) $\begin{cases} \sum_0 x_{ij} + \sum_1 x_{i\alpha} + \sum_2 x_{i\beta} = \varphi_{\alpha\beta}, \\ (\alpha, \beta) \in D, e_\alpha, e_\beta \in G^1 - T^1. \end{cases}$

其中：

$\varphi = \sum \varphi_{\alpha\beta} e_\alpha * e_\beta$ 为相应于某一 T-浸入的示嵌链.

$\sum_0$ 展开于一切无序指数偶 $(i,j) \in N, e_i \in \mathscr{P}_\beta, e_j \in \mathscr{P}_\alpha$ 之上而 $x_{ij} = x_{ji}$.

$\sum_1$ 展开于使 $(i,\alpha) \in N, e_i \in \mathscr{P}_\beta$ 的一切指数 i 上.

$\sum_2$ 展开于使 $(i,\beta) \in N, e_i \in \mathscr{P}_\alpha$ 的一切指数 i 上.

附注 方程组 (I) 将称为 G 对最大树 T(以及某一 T-浸入) 而言的**基本方程组**. 诸 $x_{ij} = x_{ji}((i,j) \in N, e_i, e_j \in T^1)$ 与 $x_{i\alpha}((i,\alpha) \in N, e_i \in T^1, e_\alpha \in G^1 - T^1)$ 将称为**基本变数**.

证. 依 §2 嵌入定理, $G \subset R^2$ 的一个必要条件为 $\varphi \sim 0$, 这里

$$\varphi = \sum \varphi_{\alpha\beta} e_\alpha * e_\beta$$

为相应于某一 T-浸入 $f : G \subset R^2, f : T \subset R^2$ 的示嵌链. 由引理 2, 这个必要条件成为: 对 $(i,j),(i,\alpha) \in D \cup N, e_i, e_j \in T^1, e_\alpha \in G^1 - T^1$, 有模 2 数 $x_{ij} = x_{ji}$ 与 $y_{i\alpha}$ 存在, 使

(2) $$\varphi = \sum x_{ij}(e_i + E_i) * (e_j + E_j) + \sum y_{i\alpha}(e_i + E_i) * e_\alpha.$$

比较两边 $e_i * e_j, e_i * e_\alpha$ 与 $e_\alpha * e_\beta(e_i, e_j \in T^1, e_\alpha, e_\beta \in G^1-T^1, (i,j),(i,\alpha),(\alpha,\beta) \in D)$ 的系数, 得

(i) $$x_{ij} = 0, \quad (i,j) \in D,$$

(ii) $$y_{i\alpha} = \sum_{e_\alpha \in \sum_j} x_{ij} = \sum_{e_j \in \mathscr{P}_\alpha} x_{ij}, \quad (i,\alpha) \in D,$$

(iii) $$\varphi_{\alpha\beta} = \sum_{e_i \in \mathscr{P}_\beta} y_{i\alpha} + \sum_{e_i \in \mathscr{P}_\alpha} y_{i\beta} + \sum_{\substack{e_i \in \mathscr{P}_\alpha \\ e_j \in \mathscr{P}_\beta}} x_{ij}, \quad (\alpha,\beta) \in D.$$

今对 $(i,\alpha) \in N, e_i \in T^1, e_\alpha \in G^1 - T^1$ 引入 $x_{i\alpha}$ 使

(iv) $$y_{i\alpha} = \sum_{e_j \in \mathscr{P}_\alpha} x_{ij} + x_{i\alpha}, \quad (i,\alpha) \in N.$$

于是对 $(\alpha,\beta) \in D, e_\alpha, e_\beta \in G^1 - T^1$ 应用 (i),(ii),(iv) 于 (iii), 即得

$$\varphi_{\alpha\beta} = \sum_{\substack{e_i \in \mathscr{P}_\beta \\ (ij) \in N}} \sum_{e_j \in \mathscr{P}_\alpha} x_{ij} + \sum_{\substack{e_i \in \mathscr{P}_\beta \\ (i\alpha) \in N}} x_{i\alpha}$$

$$+ \sum_{\substack{e_i \in \mathscr{P}_\alpha \\ (ij) \in N}} \sum_{e_j \in \mathscr{P}_\beta} x_{ij} + \sum_{\substack{e_i \in \mathscr{P}_\alpha \\ (i\beta) \in N}} x_{i\beta} + \sum_{\substack{e_i \in \mathscr{P}_\alpha \\ e_j \in \mathscr{P}_\beta \\ (ij) \in N}} x_{ij},$$

或

(v) $$\varphi_{\alpha\beta} = \sum_{\substack{e_i \in \mathscr{P}_\beta \\ e_j \in \mathscr{P}_\alpha \\ (ij) \in N}} x_{ij} + \sum_{\substack{e_i \in \mathscr{P}_\beta \\ (i\alpha) \in N}} x_{i\alpha} + \sum_{\substack{e_i \in \mathscr{P}_\alpha \\ (i\beta) \in N}} x_{i\beta}, (\alpha,\beta) \in D.$$

显然方程组 (v) 有解时, 方程组 (i) – (iv) 亦即 (2) 亦有解. 故 $G \subset R^2$ 的一个必要条件为方程组 (v) 有解. 但 (v) 式亦即 (I) 式, 故定理得证.

§4. 基本方程组的分析与简化

方程组 (I) 的分析

设 $e_\alpha, e_\beta \in G^1 - T^1, (\alpha, \beta) \in D$. (I) 中与 (e_α, e_β) 相应的方程为

$$\sum_{\substack{e_i \in \mathscr{P}_\beta \\ e_j \in \mathscr{P}_\alpha \\ (ij) \in N}} x_{ij} + \sum_{\substack{e_i \in \mathscr{P}_\beta \\ (ia) \in N}} x_{i\alpha} + \sum_{\substack{e_j \in \mathscr{P}_\alpha \\ (j\beta) \in N}} x_{j\beta} = \varphi_{\alpha\beta},$$

或简写为

(I) $$\sum\nolimits_0 + \sum\nolimits_1 + \sum\nolimits_2 = \varphi_{\alpha\beta}.$$

就 $\mathscr{P}_\alpha$ 与 $\mathscr{P}_\beta$ 的相对关系, 可分三种情形.

情形 1 $\mathscr{P}_\alpha, \mathscr{P}_\beta$ 不相遇.

此时 $\sum_0, \sum_1, \sum_2$ 都 $= 0$. 由于此时 $\varphi_{\alpha\beta}$ 必为 0, 故相应的方程必为 $0 = 0$, 方程是不足道的.

情形 2 $\mathscr{P}_a, \mathscr{P}_\beta$ 有一顶点 v 公共.

此时设 $\mathscr{P}_\alpha \cup \{e_\alpha\}$ 在 v 的两个节段为 e_a, e_b(规定在 e_a 以 v 为一端点时, e_a, e_b 中有一例如 e_b 即作为 e_a). 同样设 $\mathscr{P}_\beta \cup \{e_\beta\}$ 在 v 的两个节段为 e_r, e_s(规定在 e_β 以 v 为一端点时, e_r, e_s 中有一例如 e_s 即作为 e_β).

若 e_α, e_β 都不以 v 为端点 (见图 1), 则

$$\sum\nolimits_0 = x_{ar} + x_{as} + x_{br} + x_{bs},$$
$$\sum\nolimits_1 = \sum\nolimits_2 = 0.$$

故相应的方程为

(I)′ $$x_{ar} + x_{as} + x_{br} + x_{bs} = \varphi_{a\beta}.$$

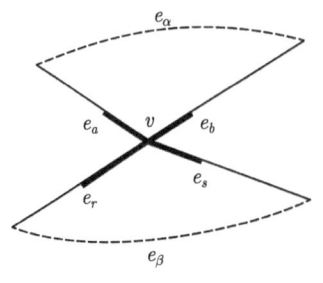 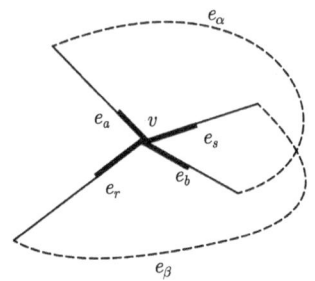

图 1

若 e_β 以 v 为端点 (因而 e_a 不再以 v 为端点),则 (见图 2)

$$\sum\nolimits_0 = x_{ar} + x_{br}, \quad \sum\nolimits_1 = 0,$$
$$\sum\nolimits_2 = x_{a\beta} + x_{b\beta} = x_{as} + x_{bs}.$$

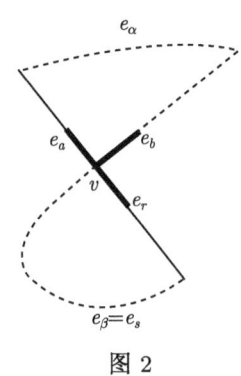

图 2

故相应的方程仍如上述形式.

在 e_a 以 v 为端点时亦然.

情形 3 $\mathscr{P}_\alpha, \mathscr{P}_\beta$ 有一以顶点 v, w 为两端的道路 $\mathscr{P} \subset T$ 公共.

设 $\mathscr{P}$ 在 v, w 处的节段为 $e_c, e_t (e_c$ 也可与 e_t 相同), $\mathscr{P}_\alpha \cup \{e_a\}$ 在 v 处的节段除 e_c 外设为 e_a(规定在 e_a 以 v 为端点时 e_a 即作为 e_a), 在 w 处的节段除 e_t 外设为 e_r (规定在 e_a 以 w 为端点时 e_r 即作为 e_a). 同样, $\mathscr{P}_\beta \cup \{e_\beta\}$ 在 v, w 处的节段除 e_c, e_t 外设为 e_b, e_s (规定在 e_β 以 v 或 w 为端点时 e_b 或 e_s 亦即为 e_β).

若 e_α, e_β 都不以 v, w 为端点 (见图 3), 则

$$\sum\nolimits_0 = x_{rs} + x_{rt} + x_{st} + x_{ab} + x_{ac} + x_{bc},$$
$$\sum\nolimits_1 = \sum\nolimits_2 = 0.$$

故相应的方程为

(I)″
$$x_{rs} + x_{rt} + x_{st} + x_{ab} + x_{ac} + x_{bc} = \varphi_{a\beta}.$$

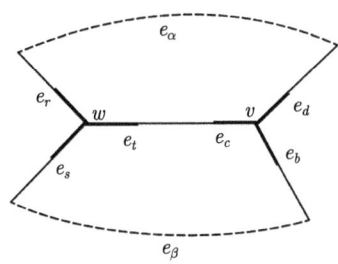 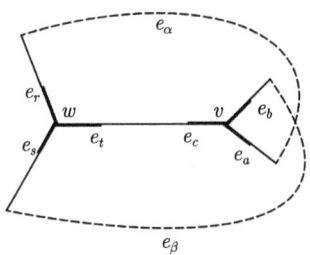

图 3

其次, 若 e_β 以 v, w 为端点 (因而 e_α 不能以 v, w 为端点), 则 (见图 4)

$$\sum\nolimits_0 = x_{rt} + x_{ac}, \quad \sum\nolimits_1 = 0,$$
$$\sum\nolimits_2 = x_{r\beta} + x_{t\beta} + x_{a\beta} + x_{c\beta}$$
$$= x_{rs} + x_{st} + x_{ab} + x_{bc}.$$

因而相应方程仍如上式 (I)″.

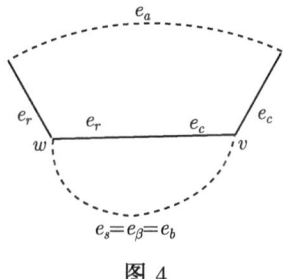

图 4

同样, 若 e_α 以 v 为端点, e_β 以 w 为端点 (见图 5), 则

$$\sum\nolimits_0 = x_{rt} + x_{bc},$$
$$\sum\nolimits_1 = x_{ba} + x_{ca} = x_{ab} + x_{ac},$$
$$\sum\nolimits_2 = x_{r\beta} + x_{t\beta} = x_{rs} + x_{st},$$

方程仍如 (I)″.

最后, 若 e_β 以 v 为端点, 而 e_α 不以 v, w 为端点 (见图 6), 则

$$\sum\nolimits_0 = x_{rs} + x_{rt} + x_{st} + x_{ac},$$

$$\sum\nolimits_1 = 0,$$
$$\sum\nolimits_2 = x_{a\beta} + x_{c\beta} = x_{ab} + x_{bc},$$

方程仍如 (I)″.

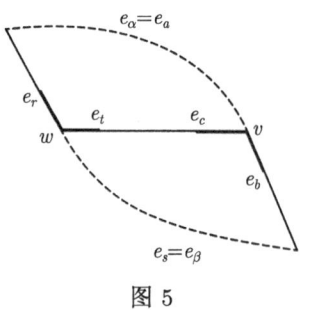
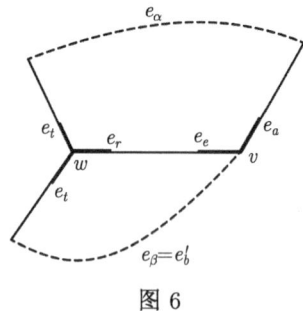

图 5　　　　　　　　　　　图 6

其余情形可从以上情形互易 e_a, e_β 或 v, w 而得出. 所得方程恒如 (I)″.

总结起来, 得

命题　方程组 (I) 的每一方程, 除不足道者 (作 $0 = 0$ 形式) 外, 必如形式

(I)′ $$x_{ar} + x_{as} + x_{br} + x_{bs} = \varphi_{\alpha\beta}$$

或

(I)″ $$x_{rs} + x_{rt} + x_{st} + x_{ab} + x_{ac} + x_{bc} = \varphi_{\alpha\beta}.$$

为了显豁起见, 特将各种情形列表如下 (两种情形可从对易例如将 e_α, e_β 或 v, w 对易而得出者不同时并列):

树根与多余变数

为了将平面图具体作出, 须要进一步将方程简化. 为此引入一些概念如下.

在最大树 T 中任取一个节梢 O, 称之为树根. 对任一顶点 $v \in G^0 = T^0$, 在 T 中有一唯一的通道连接 v 与 O, 称为到 v 的干道, 记之为 $\mathscr{P}_v$. 设 e_i 是 $\mathscr{P}_v$ 上在 v 处的节段, 则对任一在 v 处的棱 $e_j \in G^1, x_{ij}$ 将称为一多余变数. 对在 v 处而不同于 e_i 的棱 $e_j, e_k \in G^1$, 当 e_j, e_k 都 $\in G^1 - T^1$ 时, x_{jk} 称补充变数, 否则 x_{jk} 称主要变数. 此外, 对一 T-浸入 f, 除相应示嵌链

$$\varphi = \sum \varphi_{\alpha\beta} e_\alpha * e_\beta$$

中出现的一组系数 $\varphi_{\alpha\beta}, ((\alpha, \beta) \in D, e_\alpha, e_\beta \in G^1 - T^1)$ 者外, 对有一个顶点公共的 $e_\alpha, e_\beta \in G^1 - T^1$, 我们也将定义 $\varphi_{\alpha\beta}$ 为 $f(e_\alpha), f(e_\beta)$ 除去在公共端点外的交点数.

集成电路设计中的一个数学问题

情形		Σ_0	Σ_1	Σ_2	方程形式
I	(图)	0	0	0	0=0
II	(图)	$x_{ar}+x_{as}+x_{br}+x_{bs}$	0	0	(I)'
	(图)	$x_{ar}+x_{br}$	0	$x_{a\beta}+x_{b\beta}=x_{as}+x_{bs}$	(I)'
III	(图)	$x_{rs}+x_{rt}+x_{st}+x_{ab}+x_{ac}+x_{bc}$	0	0	(I)''
	(图)	$x_{rt}+x_{ac}$	0	$x_{r\beta}+x_{t\beta}+x_{a\beta}+x_{c\beta}=x_{rt}+x_{st}+x_{ab}+x_{bc}$	
	(图)	$x_{rt}+x_{bc}$	$x_{ba}+x_{ca}=x_{ab}+x_{ac}$	$x_{r\beta}+x_{t\beta}=x_{rs}+x_{st}$	
	(图)	$x_{rs}+x_{rt}+x_{st}+x_{ac}$	0	$x_{a\beta}+x_{c\beta}=x_{ab}+x_{bc}$	

扩充方程组

原来基本方程组 (I) (或 (I)′, (I)″) 中出现的变数 x_{ij}，都是主要变数或多余变数. 今设 $e_\alpha, e_\beta \in G^1 - T^1$ 而 e_α, e_β 有一公共顶点 v. 此时 $\mathscr{P}_\alpha$ 与 $\mathscr{P}_\beta$ 可只相遇于 v 亦可有一两端为 v, w 的通路 $\mathscr{P}$ 公共. 在第一情形 (见图 7) 令 $\mathscr{P}_\alpha$ 中在 v 处的节段为 e_b, $\mathscr{P}_\beta$ 中在 v 处的节段为 e_s, 又置 $e_a = e_\alpha, e_r = e_\beta$, 此时我们将引入方程

$(I)'_0$ $$x_{ar} + x_{as} + x_{br} + x_{bs} = \varphi_{\alpha\beta}.$$

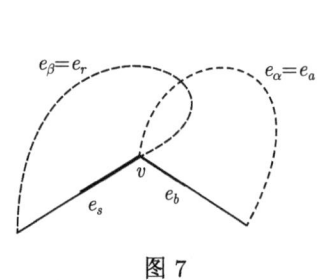 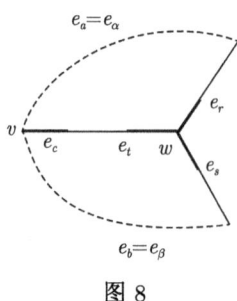

图 7　　　　　　　　　图 8

其次在第二情形 (见图 8)，我们将设 $\mathscr{P}$ 上在 v, w 处的节段各为 e_c, e_t, 在 $\mathscr{P}_\alpha$ 上 w 处不同于 e_t 的节段为 e_r，而在 $\mathscr{P}_\beta$ 上 w 处不同于 e_t 的节段为 e_s. 在 e_α 或 e_β 以 w 为端点时，则置 $e_r = e_\alpha$ 或 $e_s = e_\beta$. 又置 $e_a = e_\alpha, e_b = e_\beta$. 此时将引入方程

$(I)''_0$ $$x_{rs} + x_{rt} + x_{st} + x_{ab} + x_{ac} + x_{bc} = \varphi_{\alpha\beta}.$$

由于这些方程与原来的基本方程 (I)′, (I)″ 形式相同，且每个方程中恰含有一个补充变数 (即 x_{ar} 与 x_{ab})，而不同的方程所含的补充变数也不同. 因之，如果将这些方程加入原来的基本方程组 (I)，方程有解与否的性质将并不因之而变.

我们将称添加 $(I)'_0, (I)''_0$ 后的方程组为**扩充方程组**，并记之为 $(\tilde{I}), (\tilde{I})', (\tilde{I})''$ 等.

引理　对最大树 T 与根 O 而言，对每一节点处 $e_i, e_j \in G^1$ 引入变数 x'_{ij} 并置

$$\left.\begin{array}{l} [i,j,k] = x_{ij} + x_{ik} + x_{jk}, \\ [i,j,k]' = x'_{ij} + x'_{ik} + x'_{jk} \end{array}\right\} e_i, e_j, e_k \text{在同一节点处时}.$$

则当 x_{ij} 为已知数时，方程组

$$\begin{cases} [i,j,k]' = [i,j,k], \\ x'_{rs} = 0, \ (x_{rs}\text{为多余变数时}) \end{cases}$$

有解.

证. 在 x_{rs} 为主要变数或补充变数而 e_r, e_s 的公共节点非树根 O 时, 令自 e_r, e_s 的公共节点到 O 的通道在此节点处的节段为 e_t, 而置

$$x'_{rs} = [r, s, t],$$

在 x_{rs} 为主要变数或补充变数而 e_r, e_s 的公共节点即树根 O 时, 置 $x'_{rs} = x_{rs}$. 在 x_{rs} 为多余变数时, 则置

$$x'_{rs} = 0.$$

试证这样的 x'_{rs} 满足上述方程组. 为此, 设 $e_i, e_j, e_k \in G^1$ 的公共节点与 O 不同而自此至 O 的通道在此节点处的节段为 e_l. 若 e_l 即为 e_i, e_j, e_k 之一例如 $e_l = e_k$, 则将有

$$[i,j,k]' = [i,j,l]' = x'_{il} + x'_{jl} + x'_{ij} = x'_{ij} = [i,j,l].$$

若 e_l 不与 e_i, e_j, e_k 相同, 则将有

$$[i,j,k]' = [i,j,l]' + [i,k,l]' + [j,k,l]'$$
$$= [i,j,l] + [i,k,l] + [j,k,l]$$
$$= [i,j,k].$$

若 $e_i, e_j, e_k \in G^1$ 以 O 为公共节点则 $[i,j,k]' = [i,j,k]$ 甚显然, 故引理得证.

基本定理 II 对 G 的最大树 T 与选定的根 O 而言, 若基本方程组 (I) 可解, 则必有一解答, 它的多余变数都 $= 0$. 对于扩充方程组 (Ĩ) 亦然.

证. 已知方程组 (I) 等价于方程组 (I)′ 与 (I)″. 但 (I)′ 与 (I)″ 各可写作如下形式:

(I)′ $\qquad\qquad [ars] + [brs] = \varphi_{\alpha\beta},$

(I)″ $\qquad\qquad [rst] + [abc] = \varphi_{\alpha\beta}.$

由引理, 上方程组有解时, 下方程组也有解:

$$\begin{cases} [ars]' + [brs]' = \varphi_{\alpha\beta}, \\ [rst]' + [abc]' = \varphi_{\alpha\beta}, \\ x'_{rs} = 0, \ (x_{rs} = \text{多余变数时}) \end{cases}$$

亦即

$$\begin{cases} x'_{ar} + x'_{as} + x'_{br} + x'_{bs} = \varphi_{\alpha\beta}, \\ x'_{rs} + x'_{rt} + x'_{st} + x'_{ab} + x'_{ac} + x'_{bc} = \varphi_{\alpha\beta}, \\ x'_{rs} = 0, \ (x_{rs} = \text{多余变数时}). \end{cases}$$

或
$$\begin{cases} {\sum}'_0 + {\sum}'_1 + {\sum}'_2 = \varphi_{\alpha\beta}, \\ x'_{r_s} = 0, (x_{rs} = \text{多余变数时}) \end{cases}$$

有解, 其中 $\sum'$ 即在 $\sum^0$ 中将 x_{ij} 都易为 x'_{ij} 而得. 因之当 (I) 有解时, (I) 必有一解答它的多余变数都 $= 0$. 对 (Ĩ) 亦然, 证毕.

定义 基本方程组的一组解答将称为 T-浸入 f 的一个基本解 (对 T 与 O 而言). 同样, 扩充方程组的任意一组解答将称为 f 的一个扩充解, 多余变数都 $= O$ 的解答则将称为 f 的简化解.

附注 如果将线性图 G 改变为 G', 使对每一 $e_\alpha \in G^1 - T^1$, 在 e_α 上添入两个新顶点而分之为三段 $e_{\alpha_1}, e_{\alpha_2}, e_{\alpha_3}$, 其中 e_{α_3} 为 e_α 上中间的一段. 将 $e_{\alpha_1}, e_{\alpha_2}$ 添入 T 而成 T', 则 T' 是 G' 的一个最大树, 于是对 G', T' 而言, 方程组 (I) 与 (Ĩ) 除不足道者外将只出现图 1 与图 3 两种情形, 而不再出现图 2 与图 4-8 等情形, 若对 G' 的每一变数, 例如 $x_{a_1 i}$ 的下角出现 $\alpha_1, \alpha_2, \alpha_3$ 者都易 α_λ 为 α, 即可重行获得对 G, T 的相应方程组, 因之前面分析的许多情形实际上可以统一起来.

§5. 平面图的具体作法

我们已在 §4 中给出了 $G \subset R^2$ 的必要条件: 基本方程组 (I) 或扩充方程组 (Ĩ) 有解. 本节将证明, 这个条件也是充分的, 并将给出从方程组 (I) 或 (Ĩ) 的一个解答具体作图的方法.

调整

为了要将一具有平面性的线性图 G 在平面中具体作出, 须将一不完全是嵌入的 T-浸入逐步改变成一嵌入, 为此引入调整概念如下.

设 $f: G \subset R^2$ 是一 T-浸入, $v \in G^0 = T^0$ 的干道 $\mathscr{P}_v$ 中在 v 处的节段为 e_s[1]. 设 $e_i, e_j \in G^1$ 为在 v 处不同于 e_s 的任意两棱. 设在 f 之下, 自 e_s 起, 绕 v 作顺钟向旋转时, 在 v 处的诸棱依次为

$$e_s, e_{p_1}, \cdots, e_{p_\lambda}, e_i, e_{q_1}, \cdots, e_{q_\mu}, e_j, e_{r_1}, \cdots, e_{r_v}.$$

今另作一 T-浸入 $g: G \subset R^2$, 满足以下诸条件:

$1°$. 在 g 之下, 自 e_s 起, 绕 v 作顺钟向旋转时, 在 v 处的诸棱依次为

$$e_s, e_{p_1}, \cdots, e_{p_\lambda}, e_j, e_{q_1}, \cdots, e_{q_\mu}, e_i, e_{r_1}, \cdots, e_{r_v}.$$

---

[1] 在 $v =$ 树根 O 时, 可在 G 中添入一棱 e_0 以 O 为其一端, 于是 e_s 将取为 e_0.

$2°$. 在任一不同于 v 的顶点 w 处, 绕 w 作顺钟向旋转时, 在 w 处的诸棱在 f 下与在 g 下的次序相同.

这样的 g 必然存在, 我们将称之为将 e_i, e_j 互易的 f 的一个调整.

调整定理 (特殊形式) 设 $f: G \subset R^2$ 是一 T-浸入. 设 $e_i, e_j \in G^1$ 都在节点 v_0 处, 不位于干道 $\mathscr{P}_{v_0}$ 上, 且 e_i, e_j 在 f 下处于相邻位置. 今调整 f 为 T-浸入 g, 使 e_i, e_j 互易. 则当 f 有一组扩充解 $x_{\lambda\mu} = c_{\lambda\mu}$ 时, g 也有一组扩充解 $x_{\lambda\mu} = d_{\lambda\mu}$, 这里

$$d_{\lambda\mu} = \begin{cases} c_{\lambda\mu} + 1, & (\lambda, \mu) = (i, j), \\ c_{\lambda\mu}, & (\lambda, \mu) \neq (i, j). \end{cases}$$

再者, 若 $(c_{\lambda\mu})$ 对根 O 而言是 f 的一组简化解, 则 $(d_{\lambda\mu})$ 也是 g 的一组简化解.

证. 设 f, g 的示嵌链各为 $\varphi = \sum \varphi_{\alpha\beta} e_\alpha * e_\beta, \psi = \sum \psi_{\alpha\beta} e_\alpha * e_\beta$, 则 $(c_{\lambda\mu})$ 为下扩充方程组的解:

$$\begin{cases} (\tilde{\mathrm{I}})'_f & x_{ar} + x_{as} + x_{br} + x_{bs} = \varphi_{\alpha\beta}, \\ (\tilde{\mathrm{I}})''_f & x_{rs} + x_{rt} + x_{st} + x_{ab} + x_{ac} + x_{bc} = \varphi_{\alpha\beta}, \end{cases}$$

我们须证 $(d_{\lambda\mu})$ 为下扩充方程组的解:

$$\begin{cases} (\tilde{\mathrm{I}})'_g & x_{ar} + x_{as} + x_{br} + x_{bs} = \psi_{\alpha\beta}, \\ (\tilde{\mathrm{I}})''_g & x_{rs} + x_{rt} + x_{st} + x_{ab} + x_{ac} + x_{bc} = \psi_{\alpha\beta}. \end{cases}$$

在情形 $(\tilde{\mathrm{I}})'$ 中, 若 $v \neq v_0$ 或 $v = v_0$ 但 $(a,r),(a,s),(b,r),(b,s)$ 都 $\neq (i,j)$, 则 (e_a, e_b, e_r, e_s) 绕 v 的顺序在 f 下与在 g 下二者相同, 或至多 e_a, e_b 互易或 e_r, e_s 互易, 不论何时都有 $\psi_{\alpha\beta} = \varphi_{\alpha\beta}$. 另一面又有 $d_{ar} = c_{ar}, \cdots, d_{bs} = c_{bs}$. 因之当 $(c_{\lambda\mu})$ 满足 $(\tilde{\mathrm{I}})'_f$ 时, $(d_{\lambda\mu})$ 将满足相应的 $(\tilde{\mathrm{I}})'_g$. 反之, 若 $v = v_0$, 而在 $(a,r),(a,s),(b,r),(b,s)$ 中有一, 例如 $(a,r) = (i,j)$, 因而在 f 下绕 v 的诸棱中 e_a 与 e_r 相邻 (见图 9), 则将有 $\psi_{\alpha\beta} = \varphi_{\alpha\beta} + 1$, $d_{ar} = c_{ar} + 1$, $d_{as} = c_{as}$, $d_{br} = c_{br}$, $d_{bs} = c_{bs}$. 因而当 $(c_{\lambda\mu})$ 满足 $(\tilde{\mathrm{I}})'_f$ 时, $(d_{\lambda\mu})$ 仍将满足相应的 $(\tilde{\mathrm{I}})'_g$.

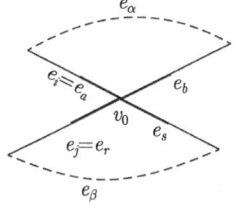
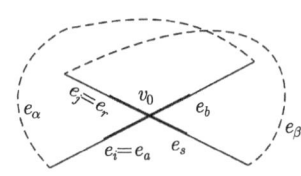

图 9

其次, 在情形 $(\tilde{I})''$ 中, 若 v, w 都 $\neq v_0$ 或 $v = v_0$ 但 $(a,b), (a,c), (b,c)$ 都 $\neq (i,j)$, 或 $w = v_0$ 但 $(r,s), (r,t), (s,t)$ 都 $\neq (i,j)$, 则显有 $\psi_{\alpha\beta} = \varphi_{\alpha\beta}, d_{rs} = c_{rs}, \cdots$. 故当 $(c_{\lambda\mu})$ 满足 $(\tilde{I})''_f$ 时, $(d_{\lambda\mu})$ 将满足相应的 $(\tilde{I})''_g$. 反之, 若 v, w 中有一, 例如 $v = v_0$, 又在 $(a,b), (a,c), (b,c)$ 中有一, 例如 $(a,c) = (i,j)$, 则将有 (见图 10)

$$\psi_{\alpha\beta} = \varphi_{\alpha\beta} + 1, \ d_{ac} = c_{ac} + 1, \ d_{ab} = c_{ab}, \cdots.$$

故 $(c_{\lambda\mu})$ 满足 $(\tilde{I})''_f$ 时, $(d_{\lambda\mu})$ 仍满足相应的 $(\tilde{I})''_g$. 其余情形亦同. 这证明了定理的前一部分.

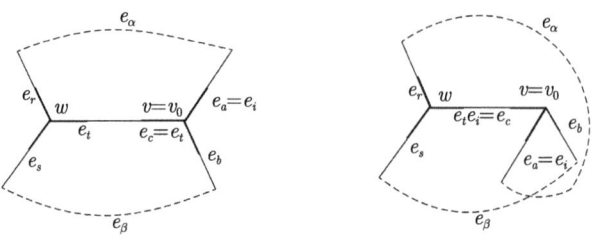

图 10

最后, 因 e_i, e_j 都不在干道 $\mathscr{P}_{v_0}$ 上, 故在简化解当 $x_{\lambda\mu}$ 为多余变数而 $c_{\lambda\mu} = 0$ 时, 同样亦有 $d_{\lambda\mu} = 0$, 因之 $(d_{\lambda\mu})$ 亦为一简化解. 证毕.

调整定理 (一般形式) 设 $f : G \subset R^2$ 是一 T-浸入, 设 $e_i, e_j \in G^1$ 在节点 v_0 处, 不位于干道 $\mathscr{P}_{v_0}$ 上, 且在 f 下 e_i, e_j 之间不含通道 $\mathscr{P}_{v_0}$ 的部分恰有 μ 个棱 $e_{q_1}, \cdots, e_{q_\mu}$ 以 v_0 为端点. 今调整 f 为 T-浸入 g, 使 e_i, e_j 互易. 则当 f 有一组扩充解 $x_{\lambda\mu} = c_{\lambda\mu}$ 时, g 也有一组扩充解 $x_{\lambda\mu} = d_{\lambda\mu}$, 这里

$$d_{\lambda\mu} = \begin{cases} c_{\lambda\mu} + 1, & (\lambda\mu) = (i,j), (i,q_1), \cdots, (i,q_\mu), (j,q_1), \cdots \text{或} (j,q_\mu), \\ c_{\lambda\mu}, & \text{在其他情形}. \end{cases}$$

再者, 若 $(c_{\lambda\mu})$ 对根 O 而言是 f 的一组简化解, 则 $(d_{\lambda\mu})$ 也是 g 的一组简化解.

证. 迭次应用特殊形式的调整定理即得.

合格化

设 $e_\alpha \in G^1 - T^1, C_a = \{e_\alpha\} \cup \mathscr{P}_a, f : G \subset R^2$ 为 T-浸入. 假设对应于 T 以及根 O 而言, f 的扩充方程组 $(\tilde{I})$ 可解且有一组简化解 $(c_{\lambda\mu})$. 如果下面的条件满足, 则称 f 对 e_α 合格而 $(c_{\lambda\mu})$ 是 f 对 e_α 的一组合格解. 条件是:

(C) 对任一 $e_i \in (C_\alpha)^t$, 以及任一与 e_i 有公共顶点的 $e_j \in G^1$, 有 $c_{ij} = 0$.

如果 f 对 e_α 不合格, 因而有 $e_i \in (C_\alpha)^1$, 以及与 e_i 有公共顶点的 $e_j \in G^1$ 使 $c_{ij} = 1$. 依据调整定理, 如果将每一对这样的 e_i, e_j 互易或改变 $(c_{\lambda\mu})$, 则可调整 f 为一 T-浸入 g, 使 g 对 e_α 合格, 且从 $(c_{\lambda\mu})$ 依调整定理所得对 g 的一组解答 $(d_{\lambda\mu})$

是 g 对 e_α 的一组合格解. 这个从 f 调整为 g 的手续将称为 f 对 e_α 的合格化. 显然, 在调整过程中可取这样的 g, 使 g 与 f 在 C_α 上相合.

引理 设 $e_\alpha \in G^1 - T^1, f: G \subset R^2$ 是一 T-浸入, 对 e_α 合格. 则可改变 f 为一 T-浸入 g, 使 $g \equiv f/T \cup \{e_\alpha\}$, 且

(i) 对任一 $\neq e_\alpha$ 的 $e_\beta \in G^1 - T^1, e_\beta$ 与 e_α 在 g 下至多在公共顶点处相遇.

(ii) 对任一 $\neq e_\alpha$ 的 $e_\beta \in G^1 - T^1, g$ 与 f 在 e_β 的两端的邻近相合.

证. 记 e_β 的两端为 a_β, b_β. 若 a_β(或 b_β) 不在 $\mathscr{P}_\alpha$ 上, 则置 $a'_\beta = a_\beta$ (或 $b'_\beta = b_\beta$). 若 a_β(或 b_β) 在 $\mathscr{P}_\alpha$ 上, 则取 e_β 内部充分邻近 a_β(或 b_β) 的一点为 a'_β(或 b'_β).

今设 $\mathscr{P}_\beta$ 不与 $\mathscr{P}_\alpha$ 相遇, 则 $a'_\beta = a_\beta$ 与 $b'_\beta = b_\beta$ 在 f 下必同在 C_α 之内或同在 C_α 之外.

若 $\mathscr{P}_\beta$ 与 $\mathscr{P}_\alpha$ 相遇于一点, 则对 e_α, e_β 的扩充方程有形式 $(\tilde{\mathrm{I}})'$. 由于 f 对 e_α 合格, 故方程左端的诸 $x_{ij} = c_{ij}$ 都 $= 0$, 因而 $\varphi_{\alpha\beta} = 0$, 故此时在 f 下 a'_β 与 b'_β(只须充分接近 a_β, b_β) 仍同在 C_α 之内或同在其外.

若 $\mathscr{P}_\beta$ 与 $\mathscr{P}_\alpha$ 相遇于一段通道, 则相应的扩充方程有形式 $(\tilde{\mathrm{I}})''$. 此时由于 f 对 e_α 合格, 方程左端的诸 $x_{ij} = c_{ij}$ 都 $= 0$, 而仍得同样结论.

故不论何时, 在 f 下 a'_β 与 b'_β 都将同在 C_α 之内或外. 因而可改变 f 在 $a'_\beta b'_\beta$ 的部分, 使在所得 g 之下 $a'_\beta b'_\beta$ 不再与 e_α 相遇, 如所欲证.

引理 设 $e_\alpha \in G^1 - T^1$, 以及 T-浸入 $f, g: G \subset R^2$ 如上一引理, 这里 f 因而 g 都对 e_α 合格. 于是 g 对 e_α 有一组合格 (简化) 解 $(c_{\lambda\mu})$ 不仅满足 (C) 且满足下述条件 $(C)^*_\alpha$ 对任意在 $(C_\alpha)^1$ 上有公共顶点的 $e_i, e_j \in G^1$, 只须 e_i, e_j 中一在 C_α 之内而一在 C_α 之外 (除公共顶点外), 即有 $c_{ij} = 0$.

证. 已知 g 有一组合格 (简化) 解 $(c'_{\lambda\mu})$ 满足条件 (C). 今依下方式改变 $(c'_{\lambda\mu})$ 为 $(c_{\lambda\mu})$: 若与 c_{ij} 对应的 e_i, e_j 如 $(C)^*_\alpha$ 中所示, 则令 $c_{ij} = 0$, 否则置 $c_{ij} = c'_{ij}$. 于是 $(c_{\lambda\mu})$ 满足条件 (C) 与 $(C)^*_\alpha$, 我们只须证明 $(c_{\lambda\mu})$ 满足 g 的扩充方程组即可.

设方程中不出现任何 x_{ij}, 此处 e_i, e_j 如 $(C)^*_\alpha$ 中所示时, 对方程中出现的诸 $x_{\lambda\mu}$ 将有 $c_{\lambda\mu} = c'_{\lambda\mu}$. 由于 $(c'_{\lambda\mu})$ 满足这一方程, 故 $(c_{\lambda\mu})$ 亦然.

其次试考虑任一对以 C_α 上一点 v 为公共顶点的 $e_i, e_j \in G^1$, 此处 e_i 在 C_α 之内而 e_j 在 C_α 之外 (除公共顶点 v 以外). 于是在 g 的扩充方程组中有 x_{ij} 的方程必如下产生: 有 $e_\beta, e_\gamma \in G^1 - T^1$ 使 $\mathscr{P}_\beta$ 含有 e_i 而 $\mathscr{P}_\gamma$ 含有 e_j. 由上一引理的证明可知在 g 下必有 e_β 在 C_α 之内而 e_γ 在 C_α 之外 (除可能在 $\mathscr{P}_\alpha$ 上的端点以外), 且 $\varphi_{\beta\gamma} = 0$.

先设 $\mathscr{P}_\beta, \mathscr{P}_\gamma$ 有一通道 $\mathscr{P}_{\beta\gamma}$ 公共. 则必有 $\mathscr{P}_{\beta\gamma} \subset \mathscr{P}_\alpha$ 且 $P_{\beta\gamma}$ 一端为 v, 而另一端设为 w. 相应的方程必如形式

$(*)$ $$x_{ij} + x_{ik} + x_{jk} + x_{rs} + x_{rt} + x_{st} = \varphi_{\beta\gamma} = 0.$$

此处 e_k, e_t 位于通道 $\mathscr{P}_{\beta\gamma}$ 上, e_r 在 $\mathscr{P}_\beta \cup \{e_\beta\}$ 上, e_s 在 $\mathscr{P}_\gamma \cup \{e_\gamma\}$ 上. e_k 以 v 为一端而 e_r, e_s, e_t 以 w 为一端. 仍由上引理, 在 g 下 e_r 在 C_α 之内或 $\mathscr{P}_\alpha$ 之上, 而 e_s 在 C_α 之外或 $\mathscr{P}_\alpha$ 之上 (端点可能除外). 由于 $(c_{\lambda\mu})$ 满足 $(C)_\alpha^*$ 故不论何时都有

$$c_{ij} = c_{ik} = c_{jk} = c_{rs} = c_{rt} = c_{st} = 0.$$

因而 $(c_{\lambda\mu})$ 满足 $(*)$.

其次设 $\mathscr{P}_\beta, \mathscr{P}_\gamma$ 只有 v 点公共, 则相应的方程形如

$$(**) \qquad x_{ij} + x_{il} + x_{kj} + x_{kl} = \varphi_{\beta\gamma} = 0,$$

此处 e_k, e_l 各在 $\mathscr{P}_\beta \cup \{e_i\}$ 与 $\mathscr{P}_\gamma \cup \{e_j\}$ 上而以 v 为端点. 与前同样不论何时都有 $c_{ij} = c_{il} + c_{kj} = c_{kl} = 0$ 因而 $(c_{\lambda\mu})$ 满足 $(**)$.

由此知 $(c_{\lambda\mu})$ 满足 g 的扩充方程组, 如所欲证.

作图法

设有线性图 G. 今取一最大树 T 与根 O. 任取一 T-浸入 $f: G \subset R^2$. 假设其相应示嵌链 $\varphi \sim 0$. 依基本定理 II, 扩充方程组 $(\tilde{I})$ 有一简化解 (c_{ij}). 今将诸 $e_\alpha \in G^1 - T^1$ 排成一次序 $e_{\alpha_1}, e_{\alpha_2}, \cdots, e_{\alpha_s}$. 依调整定理从 (c_{ij}) 出发可改变 f 为一 T-浸入 $f_1: G \subset R^2$, 使对 e_{α_1} 合格, 且由上两引理可使 f_1 满足该二引理所说的条件, 于是在 f_1 下 $e_{\alpha_2}, \cdots, e_{\alpha_s}$ 都与 e_{α_1} 至多相遇于其公共顶点, 且与 f_1 相应的扩充简化解 (c_{ij}^1) 满足条件 $(C)_{\alpha_1}^*$. 仍依调整定理从 (c_{ij}^1) 出发可改变 f_1 为一 T-浸入 $f_2: G \subset R^2$, 以及一组相应的扩充简化解 (c_{ij}^2), 使对 e_{α_2} 合格, 且满足上两引理所说的条件 (i) 与 $(C)_{\alpha_2}^*$. 由调整过程可知可取 f_2 使对 e_{α_1} 仍然合格且 (c_{ij}^2) 仍满足 $(C)_{\alpha_1}^*$. 于是在 f_2 下 $e_{\alpha_3}, \cdots, e_{\alpha_s}$ 都与 $e_{\alpha_1}, e_{\alpha_2}$ 至多相遇于其公共顶点, e_{α_1} 与 e_{α_2} 亦然. 改变 f_2 使对 e_{α_3} 合格, 且使 f_2 满足 (i), 又使相应扩充简化解 (c_{ij}^3) 满足条件 $(C)_{\alpha_3}^*$. 依次进行最后即得一 T-浸入 $f_s: G \subset R^2$ 对任一 e_{α_k} 都合格, 且任两 e_{α_k} 与 e_{α_l} 至多只相遇于其公共顶点. 这样的一个 T-浸入实为一嵌入. 因而 G 已嵌入于平面 R^2 中.

根据 §2 中嵌入定理与上作图法, 可得下面完全形式的

嵌入定理 [1] 线性图 G 具有平面性的充要条件为示嵌类

$$\Phi^2 = 0.$$

同样, §3 的基本定理 I 亦可重述之为:

基本定理 I G 有平面性的充要条件为: 对最大树 T 而言, 基本方程组 (I) 有解.

---

1) 嵌入定理的原来证明 (见 [7]) 须依据 Kuratowski 定理 ([4]), 但本文的证明并不需要.

§6. 一些估计

在以下, G 的顶点数与棱数将各设为 m 与 n.

将前数节解法在计算机上实施时, 最占存储量的是将线性方程组存入计算机的这一部分. 如果未知数的个数是 A, 则所占用计算机的字码 (bits) 数估计将为 A^2 (方程的个数并不是主要的). 如果依照 §2 所提及的方程组, A 将约略等于 $m \cdot n$, 但对于 §3-4 中所引入的简化了的基本方程组, 此数将大为减少. 而且, 我们只要估计主要变数的个数即足, 因为其余的变数可直接由主要变数表达出来而在计算机中不需要占用什么存储单元 (这些方程直接可从 G 与 T 读出).

为作这一估计, 我们将先考虑下述特殊情形:

G 的每一顶点都是三叉点, 即每一顶点处都恰有三个棱以此为端点.

今作一最大树 T, 取树根 O. 设非节梢的节点数为 r, 则节梢的个数为 $m - r$, 其中恰有 T 的三个节段以之为端点的节点数设为 s, 则恰有两个节段以之为端点的节点数将为 $r - s$. 由于 T 中节段的个数是 $m - 1$, 故有

$$m - r + 2(r - s) + 3s = 2(m - 1);$$

或

$$r + s = m - 2.$$

由于每一非节梢的节点恰恰给出一个主要变数, 而在树根处又恰有两个主要变数, 因而主要变数的个数 A 将为

$$A = r + 2 = m - s \leqslant m,$$

由此并知选择 T 时应使 T 的三叉点数 s 尽量的大, 以使 A 尽量的小.

在一般情形, 我们将易 G 为一只有三叉点的线性图 G', 使 G 中每一个 k 叉点 ($k \geqslant 3$) 易为 G' 中 $k - 2$ 个三叉点如下图所示:

$$(G) \qquad (G')$$

设 G 中无二叉点, 而 k 叉点的个数为 $m_k (k \geqslant 3)$, 则 G' 中三叉点的个数将为

$$m' = \sum (k - 2) m_k.$$

因 G 的顶点数为 m, 棱数为 n, 故有

$$m = \sum m_k,$$

$$2n = \sum km_k.$$

由此得
$$m' = 2(n - m).$$

对于 G' 的主要变数个数因之将为
$$A' \leqslant 2(n - m).$$

由于在一个电路中线路可以假定是没有电阻的, 因之与任一电路相应的线性图 G 恒可易为只有三叉点的线性图 G' 而无损于一般性. 故恒可取上面的公式来估计需解线性方程组的未知数个数, 由此以估计计算机中需占用的存储量.

§7. 杂例

例 1 Kuratowski 第一类非平面性线性图.

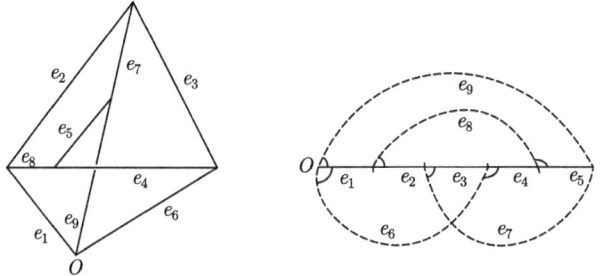

取最大树 T, 根 O 及 T-浸入 f 如上页右图. 主要变数为
$$x_{28}, x_{37}, x_{46}, x_{58}, \text{ 以及 } O \text{ 处 } x_{16}, x_{19}.$$

基本方程组 (已简化) 为
$$I_f(e_6, e_7) = 1 = x_{37} + x_{46},$$
$$I_f(e_6, e_8) = 0 = x_{28} + x_{46},$$
$$I_f(e_7, e_8) = 0 = x_{37} + x_{58},$$
$$I_f(e_8, e_9) = 0 = x_{28} + x_{58},$$
$$I_f(e_6, e_9) = 0 = x_{46} + x_{16} + x_{19},$$

以及
$$I_f(e_7, e_9) = 0 = x_{37}.$$

因方程组矛盾, 故原图形是非平面性的. 但若除去 e_9, 则相应方程组有解 (简化解) 为

$$(x_{28}, x_{37}, x_{46}, x_{58}) = (0,1,0,1), \quad 或 = (1,0,1,0).$$

而 x_{16} 任意. 今取 $x_{16} = 0$ 而从这两解进行合格化调整, 可得两个具体嵌入各如下二图所示:

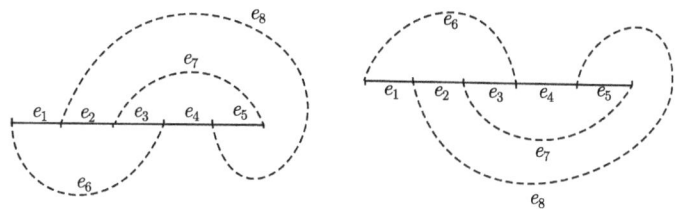

例 2 Kuratowski 第二类非平面性线性图.

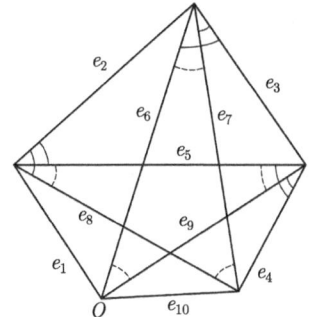

取最大树 T、树根 O, 与 T-浸入 f 如图. 已简化的基本方程组为

$$I_f(e_5, e_6) = 1 = x_{25} + x_{36},$$

$$I_f(e_5, e_7) = 1 = x_{45} + x_{37},$$

$$I_f(e_6, e_8) = 1 = x_{28} + x_{36},$$

$$I_f(e_7, e_9) = 1 = x_{37} + x_{49},$$

$$I_f(e_8, e_9) = 1 = x_{28} + x_{49},$$

$$I_f(e_5, e_{10}) = 0 = x_{25} + x_{45}.$$

由此知 e_{10} 须除去以平面化, 记所余图形为 G'. 相应基本解为

$$(x_{25}, x_{28}, x_{36}, x_{37}, x_{45}, x_{49}) = (0, 0, 1, 0, 1, 1).$$

或
$$= (1, 1, 0, 1, 0, 0).$$

而 x_{16}, x_{19} 任意.

补充变数 $x_{58}, x_{67}, x_{59}, x_{78}, x_{69}$ 可自以下补充方程求得

$$I_f(e_5, e_8) = 0 = x_{58} + x_{45},$$

$$I_f(e_6, e_7) = 0 = x_{67} + x_{36},$$

$$I_f(e_5, e_9) = 0 = x_{59} + x_{25},$$

$$I_f(e_7, e_8) = 0 = x_{78} + x_{37},$$

$$I_f(e_6, e_9) = 0 = x_{69} + x_{36} + x_{16} + x_{19}.$$

取 $x_{16} = x_{19} = 0$, 则由此所得两扩充解列表如下:

	基本变数						补充变数				
	x_{25}	x_{28}	x_{36}	x_{37}	x_{45}	x_{49}	x_{58}	x_{67}	x_{59}	x_{78}	x_{69}
第一解	0	0	1	0	1	1	1	1	0	0	1
第二解	1	1	0	1	0	0	0	0	1	1	0

依据两解经合格化调整所得图形与原图比较, 各如下所示:

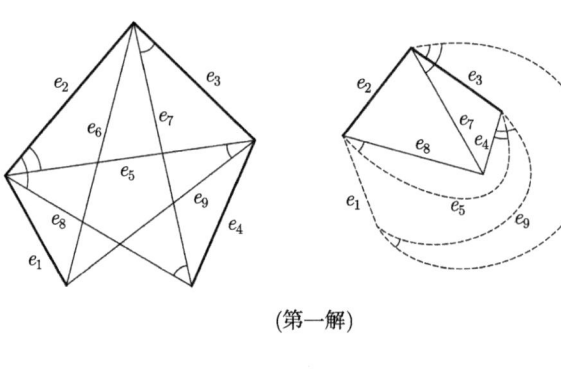

(第一解)

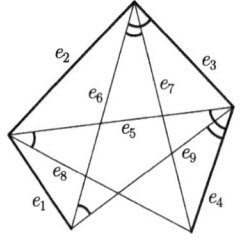
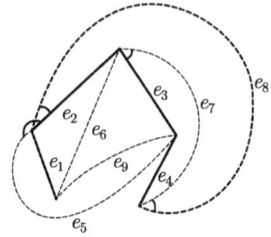

(第二解)

例 3 考虑线性图 G、最大树 T，树根 O 与 T- 浸入 f 如图：

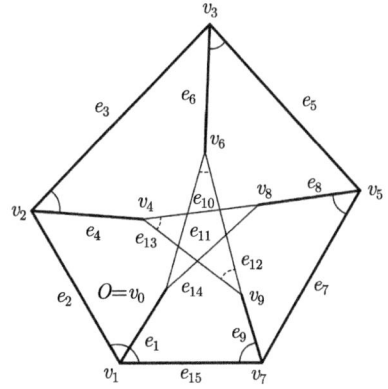

所处顶点	v_1	v_2	v_3	v_4	v_5	v_6	v_7	v_8	v_9	$O=v_0$
主要变数	$x_{2,15}$	x_{34}	x_{56}		x_{78}		$x_{9,15}$			$x_{1,11}$ $x_{1,14}$
补充变数				$x_{10,13}$		$x_{11,12}$		$x_{10,14}$	$x_{12,13}$	$x_{11,14}$

基本方程：

$$I_f(e_{10}, e_{11}) = 1 = x_{34} + x_{56},$$
$$I_f(e_{10}, e_{12}) = 1 = x_{56} + x_{78},$$
$$I_f(e_{11}, e_{13}) = 1 = x_{34} + x_{56},$$
$$I_f(e_{12}, e_{14}) = 1 = x_{56} + x_{78},$$
$$I_f(e_{13}, e_{14}) = 1 = x_{34} + x_{78}.$$

由此知 e_{14} 应从 G 中除去，而最后两方程可以略去，

$$I_f(e_{11}, e_{15}) = 0 = x_{56} + x_{2,15},$$
$$I_f(e_{14}, e_{15}) = 0 = x_{78} + x_{2,15},$$

由此知 e_{15} 应从 G 中除去。

命除去 e_{14}, e_{15} 后的图形为 G'。此时主要变数只有 x_{34}, x_{56}, x_{78} 以及 $x_{1,11}$。基本方程只有两个：

$$\begin{cases} x_{34} + x_{56} = 1, \\ x_{56} + x_{78} = 1. \end{cases}$$

又此时补充变数也只有三个：$x_{10,13}, x_{11,12}, x_{12,13}$。补充方程为

$$\begin{cases} I_f(e_{10}, e_{13}) = 0 = x_{10,13} + x_{78}, \\ I_f(e_{11}, e_{12}) = 0 = x_{11,12} + x_{56}, \\ I_f(e_{12}, e_{13}) = 0 = x_{12,13} + x_{56}. \end{cases}$$

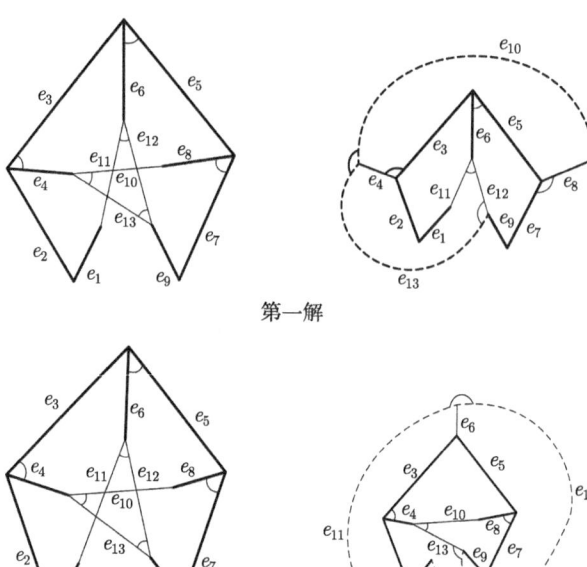

第一解

第二解

	主要变数			补充变数		
	x_{34}	x_{56}	x_{78}	$x_{10,13}$	$x_{11,12}$	$x_{12,13}$
第一解	1	0	1	1	0	0
第二解	0	1	0	0	1	1

($x_{1,11}$ 任意, 可取为 0.)

参考资料

[1] L.Auslander-S. V. Parter. On imbedding graphs in the sphere. J. Math. Mech., 1961, 10: 517-523.

[2] W. Bader. Topologische Problem der gedruckten Schaltung und seine Lösung. Archiv für Elektrotechnik, 1964, 49: 2-12.

[3] G. J. Fisher-O. Wing. Computer recognition and extraction of planar graphs from the incidence matrix. IEEE: Trans. on Cricuit Theory, 1966, CT-13: 154-163.

[4] C. Kuratowski. Sur le problème des courbes gauches en topologie. Fund. Math., 1930, 15: 271-283.

[5] S. MacLane. A combinatorial condition for planar graphs. Fund.Math., 1937, 28: 22-32.
[6] H. Whitney. Non-separable and planar graphs. Trans. Amer. Math. Soc., 1932, 34: 339-362.
[7] 吴文俊. 复合形在欧氏空间中的实现问题, Ⅰ, Ⅲ. 数学学报, 1955, 5: 505-552; 1958, 8: 79-94.

线性图的平面嵌入 *

设 G 是连通线性图. 关于 G 在平面 R^2 中的嵌入问题, 早在 30 年代即已由 Kuratowski, Whitney, MacLane 等给出了多种不同的判准而获得理论上的解决. 60 年代以来由于实际应用的需要转而讨论 G 可嵌入 R^2 中的具体作法的文献较多. 本文对 G 在 R^2 中的嵌入给出了另一种判准, 据此并可将不同嵌入进行分类且容易得出具体作图方法.

先说明一些符号与概念. 在 G 可嵌入平面 R^2 中时, 对任两嵌入 $f,g:G \subset R^2$, 若有 R^2 到自身保持定向的拓扑变换 φ, 使 $\varphi f = g$, 则称 f,g 等价, 记作 $f \sim g$, G 到 R^2 中嵌入的等价类记作 $I(G)$. 若 T 是 G 的一个子图, 对嵌入 $f,g:G \subset R^2$ 有 $f/T \sim g/T$, 则称 f,g 对 T 等价, 记作 $f_{\widetilde{T}} g$, G 到 R^2 中嵌入的 T 等价类记作 $I_T(G)$.

在以下 T 将是 G 的一个固定的通过所有顶点的树. 我们的目的在于确定 $I_T(G)$ 与 $I(G)$.

记 G,T 的棱的集合为 G^1, T^1, 顶点的集合为 $G^0 = T^0$, T 的自由端的集合为 T^{00}, $T^0 - T^{00}$ 为 T^*. 我们将假定: G^1 中无一棱两端相同, 无两棱有相同顶点, T^{00} 中有一固定的自由端 O, 在 $G^t - T^1$ 中无一棱以 O 或 T^* 中的顶点为端点, 而对 T^{00} 中任一不是 O 的自由端, 在 $G^t - T^1$ 中恰有一棱以此为端点. 由于我们恒可将原线性图转化为满足以上条件的图来讨论, 这些假定并无损于一般性.

记 $T^1 = \{e_s/s \in J\}, G^1 - T^1 = \{e_\alpha/\alpha \in A\}$. 对于任意 $v, w \in T^0$, v 与 w 间在 T 上有唯一通道记作 P_{vw}, 特别在 $W = O$ 时, 记 P_{vw} 为 P_v. 在 P_v 上以 v 为端点的棱记作 e_v. 对任一 $e_s \in T^1$, 以 $V(e_s)$ 表 e_s 的这一端点使 $P_{V(e_s)}$ 与 e_s 只有 $V(e_s)$ 公共. 置 $E(v) = \{e_s/V(e_s) = v, e_s \neq e_v\}$. 又以 N_v 表一切无序偶 $(r,s) = (s,r)$ 的集合, 以 $\widetilde{N}_v$ 表一切有序偶 $[r,s]$ 的集合, 这里 $e_r, e_s \in E(v)$. 记 $\bigcup_{v \in T^0} N_v = N$, $\bigcup_{v \in T^0} \widetilde{N}_v = \widetilde{N}$. 对于 $e_\alpha \in G^1 - Y^1$, 在 T 中有唯一通道联结 e_α 的两端, 记作 P_α, 在 P_α 上有唯一顶点 v_α, 使 P_{v_α} 与 P_α 只有 v_α 公共.

以 Z_2 表模 2 数域. 以下所有的数都属于 Z_2. 今对任意嵌入 $f:T \subset R^2$ 引入一些 Z_2 中的数如下: 对任意 $[i,j] \in \widetilde{N}_v, v \in T^*$, 若在 f 下 e_v, e_i, e_j 依反时针方向旋转, 则置 $f_{ij} = 0$, 否则置 $f_{ij} = 1$. 若 $a, b \in T^{00}, P_a, P_b$ 以 P_v 为公共通道, $v \in T^*$, 在 P_{v_a}, P_{v_b} 上以 v 为端点的棱各为 e_i, e_j, 则以 f_{ab} 表 f_{ij}. 若 $e_\alpha \in G^1 - T^1, e_\alpha$ 的两端为 a, a', 而 $b \in T^{00}, b \neq a, a'$, 则置 $f_{ba} = f_{ba} + f_{ba'}$. 若 $e_\beta \in G^1 - T^1, \beta \neq \alpha, e_\beta$

---

* 本文原载《科学通报》, 1974 年第 5 期.

的两端为 b, b' 则置 $f_{\beta\alpha} = f_{ba} + f_{b'a}$, 显然有

$$f_{ij} = f_{ji} + 1, \quad f_{ab} = f_{ba} + 1, \quad f_{\beta\alpha} = f_{\alpha\beta}.$$

若将 $f: T \subset R^2$ 任意拓广为一浸入 $\tilde{f}: G \subset R^2$, 使在 $\tilde{f}$ 下任一 $e_\alpha \in G^1 - T^1$ 除端点外都不与 T 相遇 (这时 $\tilde{f}$ 简称为一 T-浸入), 则 $f_{\alpha\beta} = f_{\beta\alpha}$, 即 $\tilde{f}(e_\alpha)$ 与 $\tilde{f}(e_\beta)$ 的 (模 2) 相交指数, 且 $f_{ba} = \text{Ord}_b \tilde{f}(P_\alpha \cup e_\alpha) + \text{Ord}_o \tilde{f}(P_\alpha \cup e_\alpha)$, 此处 $\text{Ord}_p C$ 是指 R^2 中一点 p 对一简单闭曲线 C 的 (模 2) 次数, 其值为 0 或 1, 视 p 在 C 之外或内而定.

对任意 $(i, j) \in N$ 引入未知数 $x_{ij} = x_{ji}$, 与 $f_{ab}, f_{\alpha\beta}, f_{\beta\alpha}$ 一样从 x_{ij} 定义 $x_{ab}, x_{\alpha\beta}, x_{\beta\alpha}$. 但此时

$$x_{ij} = x_{ji}, \quad x_{ab} = x_{ba}, \quad x_{\beta\alpha} = x_{\alpha\beta}.$$

又对任意 $\alpha \in A$ 引入变数 y_α. 作 (模 2) 方程组:

$(\text{I})_f \; x_{\alpha\beta} = f_{\alpha\beta} (\alpha, \beta \in A, \alpha \neq \beta),$

$(\text{II})_f \; (f_{ij} + x_{ij})(f_{i\kappa} + x_{i\kappa}) + (f_{i\kappa} + x_{i\kappa})$
$\quad \times (f_{ji} + x_{ji}) + (f_{\kappa i} + x_{\kappa i})(f_{\kappa i} + x_{\kappa i}) = 1$
$\quad (e_i, e_j, e_\kappa \in E(v), v \in T^*, i \neq j \neq \kappa \neq i),$

$(\text{III})_f \; y_\alpha y_\beta + (x_{a\beta} + f_{a\beta}) y_\alpha + (x_{b\alpha} + f_{b\alpha}) y_\beta = 0$
$\quad (\alpha, \beta \in A, \alpha \neq \beta, a, b)$ 各为 e_α, e_β 的任一端).

我们的主要结果如下 ($f: T \subset R^2$ 是一固定嵌入):

定理 1 方程组 $(\text{I})_f$ 有解时, 方程组 $(\text{I})_f, (\text{II})_f$ 也有解.

定理 2 $I_T(G)$ 与方程组 $(\text{I})_f, (\text{II})_f$ 的解一一对应.

定理 3 $I(G)$ 与方程组 $(\text{I})_f, (\text{II})_f, (\text{III})_f$ 的解一一对应.

这些定理的证明简述如下.

对嵌入 $f: T \subset R^2$ 记 $(\text{I})_f$ 的一切解的集合为 S_f, 记 $c = (c_{rs}) \in S_f$, 即 $x_{rs} = c_{rs}$ 满足 $(\text{I})_f$. 设 $g: T \subset R^2$ 是另一嵌入. 对任意 $(r, s) \in N_v, v \in T^*$, 置 $d_{rs} = c_{rs}$ 或 $c_{rs} + 1$, 视 (e_v, e_r, e_s) 在 f 与 g 下有同一旋向与否而定, 则 $d = (d_{rs}) \in S_g$, 此时记为 $(f, c) \sim (g, d)$.

引理 1 设 $f: T \subset R^2$, 而 $c = (c_{rs}) \in S_f$ 满足 $(\text{II})_f$ 中对于某一 $v \in T^*$ 的一切方程, 则必有 $(g, d) \sim (f, c)$, 使对任意 $(r, s) \in N_v$ 有 $d_{rs} = 0$, 而对 $(r, s) \in N_w, w \neq v$, 有 $d_{rs} = Crs, g_{rs} = f_{rs}$.

证. 对任意 $e_i, e_j, e_\kappa \in E(v), i \neq j \neq \kappa \neq i, c$ 满足 $(\text{II})_f$ 中相应方程的条件为 $f_{ij} + c_{ij}, f_{j\kappa} + c_{j\kappa}, f_{\kappa i} + c_{\kappa i}$ 不能全为 0 或全为 1, 或在 f 下 e_v, e_t 等依反时针方

向次序排列为例, 如 e_v, e_i, e_j, e_κ 时, $(c_{ij}, c_{j\kappa}, c_{i\kappa})$ 不能取值 (0,0,1) 或 (1,1,0). 由此知可取 $g: T \subset R^2$, 使在 g 下 e_i, e_i, e_κ 依反时针方向重行排列次序 $e_v, e_{i'}, e_{j'}, e_{\kappa'}$, 而 $(g,d) \sim (f,c)$ 时, 将有 $d_{i'j'} = d_{i'\kappa'} = d_{j'\kappa'} = 0$, 亦即 $d_{ij} = d_i\kappa = d_j k = 0$. 由此几何意义对 $E(v)$ 中的棱数进行归纳法即得引理.

定理 2 的证明 设 $g: G \subset R^2$, 则对 $(r,s) \in N$ 置 $d_{rs} = 0$ 时, $d = (d_{rs})$ 显然满足 $(I)_g$ 与 $(II)_g$. 命 $(f,c) \sim (g,d)$, 则 c 满足 $(I)_f$ 与 $(II)_f$. 故任一 $g: G \in R^2$ 对应于 $(I)_f$ 与 $(II)_f$ 的一组解. 反之, 若 c 是 $(I)_f$ 与 $(II)_f$ 的一组解, 则由引理 1 可得 $(g,d) \sim (f,c)$, 此处 $g: T \in R^2$ 而 $d = (d_{rs}), d_{rs} = 0$. 由 $(I)_g, g_{rs} = 0$ 因而可将 $g: T \subset R^2$ 扩充为一嵌入 $g: G \subset R^2$. 若 C 与 C' 是 $(I)_f, (II)_f$ 的两组不同解答, 因而有 $(r,s) \in N_r, v \in T^*$, 使 $c_{rs} \neq c'_{rs}$, 则在与 c, c' 相应的 $g, g': T \subset R^2$ 下, e_v, e_r, e_s 将有不用旋向而有 $g\tilde{T}g'$. 故 $I_t(G)$ 与 $(I)_t, (II)_t$ 的解一一对应.

引理 2 设 $(I)_f$ 有解而 $c = (c_{rs}) \in S_f$, 又 $e_\alpha \in G^1 - T^1$, 则有 $(g,d) \sim (f,c)$ 以及 $d' = (d'_{rs}) \in S_g$ 满足以下条件:

1. 对任一 $e_\beta \in G^1 - T^1, \beta \neq \alpha$, 有 $g_{\alpha\beta} = 0$.
2. 对任一 $e_\kappa \subset P_\alpha$, 以及 $e_r, e_s \in EV(e_\kappa), e_r, e_s$ 不同于 $V(e_k)$ 的端点为 a, b, 则只须 $g_{a\alpha} \neq g_{b\alpha}$, 即有 $d'_{rs} = 0$.
3. 对任一 $e_\kappa \subset P_\alpha$ 以及 $e_s \in EV(e_\kappa), e_s \neq e_\kappa$, 有 $d'_{s\kappa} = 0$.

此时将称 (g, d') 对 e_a 合格.

证. 从 f 改变诸棱在 R^2 中的旋向可得 $(g,d) \sim (f,c)$, $g: T \subset R^2$ 与 $d = (d_{rs}) \in S_g$ 使以下诸条件满足:

(1) 记 P_a 上 $u = u_a$ 处的两棱为 e_j, e_l, 有 $g_{ij} = 0, d_{lj} = 0$.

(2) 设 $e_r \in E(v), e_r \neq e_l, e_i$. 若 c 满足 $(II)_f$ 中与 (e_r, e_i, e_j) 相应的方程, 则有 $d_{ri} = d_{ri} = 0$; 若 C 不满足 $(II)_f$ 中与 (e_r, e_i, e_j) 相应的方程, 则有 $g_{ir} = g_{ri} = 0, d_{ri} = d_{ri} = 1$.

(3) 对任一 $e_\kappa \subset P_a, V(e_\kappa) = v' \neq v$, 以及 $e_s \in E(v'), e_s \neq e_\kappa$, 有 $d_{sk} = 0$

易见 g 满足引理 2 的第一个条件.

其次, 记 $v = v_a$ 与 e_i, e_j 如前而改变 $d = (d_{rs})$ 为 $d' = (d'_{rs})$ 如下:

$(1)'$ 若 $e_r \in E(v), e_r \neq e_i, e_j, c$ 不满足 $(II)_f$ 中, 因而 d 不满足 $(II)_g$ 中与 (e_r, e_i, e_j) 相应的方程 (见条件 (2)), 则置 $d'_{ri} = d_{ri} + 1, d'_{rj} = d_{ri} + 1$, 即 $d'_{ri} = d'_{rj} = 0$.

$(2)'$ 若 $e_r, e_s \in E(u), g_{ir} = g_{ri} = 0, g_{is} = g_{sj} = 0$, 而 $(II)_f$ 中与 (e_r, e_i, e_j) 以及 (e_s, e_i, e_j) 相应的两方程中有一为 c 所满足而另一则不满足, 亦即 $(II)_g$ 中的两个相应方程中一为 d 所满足而另一则否, 则置 $d'_{rs} = d_{rs} + 1$.

$(3)'$ 若 $e_\kappa \subset P_a, e_r, e_s \in EV(e_\kappa), g_{r\kappa} = g_{\kappa s}, d_{rs} = 1$, 则置 $d'_{rs} = d_{rs} + 1$, 即 $d'_{rs} = 0$.

$(4)'$ 在其他情形, 都置 $d'_{rs} = d_{rs}$

由 (1)′, (2)′ 可得以下性质:

(5)′ 记 $E^*(v)$ 为 e_i, e_j 以及使 $g_{ir} = g_{ri} = 0$ 的一切 $e_r \in E(v)$ 的集合, 则对任意互不相同的 $e_r, e_s, e_i \in E^*(v)$, 在 $d'_{rs}, d'_{rl}, d'_{si}$ 中各与 d_{rs}, d_{rt}, d_{st} 不相等的个数为 0 或 2.

由以上可知对任意 $e_\beta, e\gamma \in G^1 - T^1, \beta \neq \gamma$, 有 $d_{\beta\gamma} = d'_{\beta\gamma}$ 定义如 $f_{\beta\gamma}, x_{\beta\gamma}$ 等), 因而 d' 与 d 同样满足 $(I)_g$ 中的方程 $x_{\beta r} = g_{\beta r}$, 即 $d' = (d'_{rs})$ 与 d 同样仍为 $(I)_g$ 的解, 且 (g, d') 满足引理中的诸条件.

定理 1 的证明 设 $v \in T^*$. 可设 G 不能分解成只有 v 点公共的两个子图的和 (一般情形容易归结于此). 于是可将 $v_a = v$ 的诸 $e_a \in G^1 - T^1$ 排成一次序 $e_{a_1}, \cdots, e_{a_k}$, 使每一 P_{a_i} 必与 $P_{a_1}, \cdots, p_{a_{i-1}}$ 之一有公共棱. 今依引理 2 的作法由 $(g_0, d_0) = (f, c)$ 出发依次作 $(g_i, d_i) \sim (g_{i-1}, d'_{i-1})$ 以及 $d' \in S_{gi}(d'_0 = d_0)$, 使 (g_i, d'_i) 对 e_{a_i} 合格, 则由作法可知 $(g, d') = (g_k, d'_k)$ 对所有 $e_{a_1}, \cdots, e_{a_k}$ 合格. 特别有 $d'_{rs} = 0, (r, s) \in N_v$, 而 d' 满足 $(II)_g$ 中在 v 处的所有方程. 对每一 $v \in T^*$ 取同样作法即得一 $g^* : T \subset R^2$ 与 $d^* = (d^*_{rs}) \in S_{g*}$ 而 $d^*_{rs} = 0, (r, s) \in N$. 取 $(f, c^*) \sim (g^*, d^*)$, 则 c^* 满足 $(I)_f, (II)_f$.

定理 3 的证明 设 $g : G \subset R^2, d = (d_{rs}) \in S_g, d_{rs} = 0$. 命 $(f, c) \sim (g, d), f : T \subset R^2, c$ 满足 $(I)_f$ 与 $(II)_f$. 置 $u_a = \text{Ord}_o g(P_a \cup e_a)$, 则 $y_a = u_a$ 与 $x_{rs} = C_{rs} = 0$ 满足 $(III)_g$. 因而 $y_a = u_a$ 与 $x_{rs} = c_{rs}$ 满足 $(III)_f$, 即任一嵌入 $g : G \subset R^2$ 必与 $(I)_f, (II)_f, (III)_f$ 的一组解答对应. 反之, 设 $x_{rs} = C_{rs}, y_a = u_a$ 是 $(I)_f, (II)_f, (III)_f$ 的一组解答, 则由定理 2 取 $g : T \subset R^2$ 与 $(g, d) \sim (f, c), d = (d_{rs}), d_{rs} = 0$ 时, $x_{rs} = d_{rs}, y_a = u_a$ 将满足 $(I)_g, (II)_g, (III)_g$. 由于 $(III)_g$ 可写作 $y_a(y_\beta + g_{a\beta}) = g_{ba} y_\beta$, 故将 $G^1 - T^1$ 中诸棱排成一定次序后, 将 $g : T \subset R^2$ 逐步拓广为 $g : G \subset R^2$ 使 $\text{Ord}_o g(P_a \cup e_a) = u_a$. 故知 $I(G)$ 与 $(I)_f, (II)_f, (III)_f$ 的解一一对应.

On the Planar Imbedding of Linear Graphs*

1. Introduction

The present paper is a reproduction of the results already published in Chinese from 1973 onwards. It is concerned with the problem of planar imbedding of linear graphs (supposed to be connected and possessing no loops henceforth). The problem may be separated into four parts:

P1. Decide whether a connected linear graph (or *graph* for short) G is imbeddable in the plane (or *imbeddable* for short).

P2. Decide, in the case of a non-imbeddable graph. G, a minimal set of edges the removal of which will render the remaining part of G imbeddable.

P3. Give a method of imbedding G in the plane in the case G is imbeddable.

P4. Give a description of the totality of possible imbeddings of G in the plane in the case G is imbeddable.

The problem P1 was already solved in the early thirties. Thus, Kuratowski has given the following simple and elegant criterion [KU1]: Let $K1$ be the graph with five vertices and all edges connecting any two of them. Let $K2$ be the graph with two triads of vertices and all edges connecting pairs of vertices one from each triad. Then we have the following.

Theorem of Kuratowski *A graph G is imbeddable if and only if it does not contain any subgraph of type $K1$ or $K2$.*

Similar criteria have been given by Whitney and MacLane, also in the thirties. However, all these criteria are only *existential* in character, although they settle the problem P1 quite satisfactorily at least in a theoretical sense. In fact, these criteria give no means of a *constructive* manner for deciding whether a graph concretely given is planar or not. For example, for the Kuratowski criterion we have no means of detecting subgraphs of type $K1$ or $K2$ well hidden in a concretely given graph. This

---

*本文原载《系统科学与数学》, 1985, 5(4): 290-302.

fact thus has deprived these criteria of any practical value.

After more than twenty years of silence the interest in the problem revived in the early sixties owing seemingly to practical needs. This time however, the interest lay no more on theoretical imbeddability of a linear graph, but rather, on practical decision of the imbeddability of any given graph in giving algorithmic procedures. Beginning from a paper by Auslander and Parter [AP1], the study culminated in a paper of Hopcroft and Tarjan [HT1] in giving an efficient planarity algorithm for a linear graph. Nevertheless their method gives merely an answer to problem P1 from the practical side and leaves problems P2–4 completely untouched. As mentioned by the authors themselves in their joint paper, their "planarity algorithm$\cdots$tests a graph G for planarity, but it does not actually construct a planar representation of G." We remark that it is just the latter part corresponding to problems P2–3 above that renders the study of planarity of graphs so important in applications.

On the other hand the present author discovered in 1967 a solution to problem P1 which is both of theoretical interest and of practical value in being algorithmic. The method was based on a theory of imbedding and immersion of complexes in a Euclidean space [WU1] and was applied this time to linear graphs, i. e. complexes of dimension 1. It leads to the criterion that a graph is imbeddable if and only if a certain system of linear equations on mod 2 coefficients is solvable in integers mod 2. These results, owing to circumstances, were not published until late 1973, cf. [WU2]. Now each equation in the linear system of our criterion has either two or four variables. In 1978 Liu Yan-pei made an important complement to our methd in reducing each such equation to one with only two variables [L1]. This enables the decision of planarity to be carried out actually without any computation and is extremely feasible. However, in either [HT1] or [L1] or [TU1] only criteria of imbeddability were given, with the important problem of actual imbedding in the case the graph is imbeddable entirely untouched. In the meantime the present author arrived at a complete solution of all problems P1–4 listed above and the proofs were purely algebraic with no more use of algebraic topology. These results were published as an appendix to the Chinese version of the book [WU1], cf. [WU4].

The present paper has the aim of giving an English version of all these results, so far published only in Chinese, with due modifications.

To fix the ideas, throughout the paper the following notations will be adopted:

We will always work over integers mod 2 and the field of mod 2 integers will be denoted as usual by $Z2$.

The plane in which graphs are to be imbedded is denoted by $R2$.

The graph (connected without loops) is denoted by G, with numbers of vertices Nv and number of edges Ne.

The vertices of G are Vi, with i running over some index set I. The collection of all such vertices will be denoted by $V(G)$, or simply V.

The edges of G are Eq, with q running over some index set Q. The collection of all such edges will be denoted by $E(G)$, or simply E.

The letters $i, j, k, l, \cdots$ will be used for indices in I, and the letters $q, r, s, t, \cdots$ for those in E.

The set of all unordered pairs of edges (Er, Es), with Er, Es disjoint from each other, will be denoted by $D2(G)$, or simply $D2$.

The set of all pairs (Vi, Eq), with Vi not an end of Eq, will be denoted by $D1(G)$, or simply $D1$.

The collection of all functions

$$A: D1 \to Z2$$

forms naturally an additive group and will be denoted by $C1(G)$, or simply $C1$.

The collection of all functions

$$F: D2 \to Z2$$

forms naturally an additive group and will be denoted by $C2(G)$, or simply $C2$.

For any pair (Vi, Eq) in $D1$, the function A in $C1$ which takes the value 1 on (Vi, Eq) but 0 on any other pair in $D1$ will be denoted by $\langle Vi, Eq \rangle$.

Similarly, for any pair (Eq, Er) in $D2$, the function F in $C2$ which takes the value 1 on (Eq, Er) but 0 on any other pair in $D2$ will be denoted by $\langle Eq, Er \rangle$.

The morphism

$$d: C1(G) \to C2(G)$$

defined by

$$dA/(Eq, Er) = A/(Vi, Er) + A/(Vj, Er) + A/(Vk, Eq) + A/(Vl, Eq)$$

for

$$A \text{ in } C1, \quad Eq = ViVj, \quad Er = VkVl, \quad (Fq, Er) \text{ in } D2,$$

will be called the *differential* in G.

For any two broken lines $L1$, $L2$ in $R2$ not both closed to become polygons for which the ends of $L1$ (resp. $L2$), which exist if not closed, are disjoint from $L2$ (resp. $L1$), there is a well-defined *intersection number* in $Z2$ which will be denoted by $\text{Int}(L1, L2)$.

For any closed polygon P with possibly self-intersections and a point A not on P there is the well defined *order* of A with respect to P in $Z2$ which will be denoted by $\text{Ord}(A, P)$.

If B is another point in $R2$ not on P and L is a broken line joining A and B, we would have the following relation in $Z2$:

$$\text{Ord}(A, P) + \text{Ord}(B, P) = \text{Int}(L, P).$$

If A, B, P are as above with P a simple closed polygon and

$$\text{Ord}(A, P) = \text{Ord}(B, P)$$

in $Z2$, then by the theorem of Jordan A, B can be joined by a simple broken line in $R2$ disjoint from P.

2. A criterion for imbeddability

Without loss of generality we shall restrict maps of G in the plane $R2$ to piecewise linear ones which will always be so assumed in what follows. A (piecewise linear) map

$$f: G \to R2$$

is called an *imbedding* if it is topological or 1-1.

Let H be any subgraph of G. Then a map f will be called an *H-immersion* of G if the following conditions (a) — (e) are observed:

(a) The images of vertices are all different.

(b) The image of each edge is a simple broken line.

(c) The image of any vertex is not on the image of any edge except at the possible end.

(d) f is an imbedding when restricted on H, while for any edge Eq not in H, $f(Eq)$ will not meet $f(H)$ except possibly at vertices common to H and Eq.

(e) The images of any two edges will meet at most at a finite number of points besides the possible common ends.

The H-immersion will simply be called an *immersion* of G if H is the empty sub-graph.

Consider now any immersion

$$f : G \to R2.$$

Definition The element $c(f)$ in $C2(G) = C2$ defined by

$$c(f)/(Eq, Er) = \text{Int}(fEq, fEr), \quad \text{for } (Eq, Er) \text{ in } D2,$$

will be called the *immersion element* defined by f.

Theorem 1 *For any two immersions f and g of G in the Plane*

$$f, g : G \to R2,$$

the elements $c(f)$, $c(g)$ belong to the same class of the quotient group $C2/dC1$.

Proof. Consider first the case f and g coincide on all vertices of G and all edges of G except a single one, say the edge Es.

Now fEs, gEs form a polygon P (with possibly self-intersections). For any vertex V_k of G disjoint from Es let us set

$$Ok = \text{Ord}(fVk, P), \quad \text{for } Vk \text{ disjoint from } Es.$$

Define now an element c in $C1$ by

$$c = \text{SUM } Ok.\langle Vk, Es\rangle,$$

the summation being over all vertices Vk disjoint from Es.

Now for any edge Eq disjoint from Es we have $fEq = gEq$. Therefore, with Vk, Vl as the two ends of Eq, we would have

$$c(f)/(Eq, Es) + c(g)/(Eq, Es) = \text{Int}(fEq, fEs) + \text{Int}(gEq, gEs)$$
$$= \text{Int}(fEq, P) = \text{Ord}(fVk, P) + \text{Ord}(fVl, P).$$

On the other hand if the ends of Es are Vi and Vj, then we would have

$$dc/(Eq, Es) = c/(Vi, Eq) + c/(Vj, Eq) + c/(Vk, Es) + c/(Vl, Es)$$
$$= Ok + Ol.$$

Comparing, we have

$$c(f) + c(g) = dc/(Eq, Es).$$

For any pair (Eq, Er) in $D2$ with Eq, Er both different from Es, it is clear that

$$c(f) + c(g) = dc = 0/(Eq, Er).$$

So the assertion is proved in the above case.

Consider now the case that f and g coincide on all vertices of G and are arbitrary otherwise. We may always define a sequence of immersions $h0, h1, \cdots, hs$ such that $h0$ coincides with f, hs with g, and each hr coincides with the preceding one with the exception of a single edge. By the preceding case each element $c(hr)$ will belong to the same class of $C2/dC$ 1 as the element c of the preceding immersion h in the sequence. It follows that the assertion still holds true in the present case.

Consider now the general case with f, g arbitrary, with however the images of all vertices under both f and g different from each other.

For any vertex Vi let us draw a simple broken line Li in the plane with ends fVi and gVi such that, what is clearly possible, these broken lines are mutually disjoint.

For each edge Eq let us join now the ends of fEq by a broken line Lq disjoint from all Li except possibly at the images of their common ends.

Define now an immersion h and an immersion h' of G by

$$h(Eq) = Lq,$$
$$h'(Eq) = Li + Lq + Lj,$$

where Vi, Vj are the two ends of Eq.

From the construction we see that the elements $c(h)$ and $c(h')$ belong to the same class of $C2/dC$ 1. On the other hand by the preceding cases already proved $c(h)$ and $c(h')$ are in the same classes of $C2/dC$ 1 as $c(f)$ and $c(g)$ respectively. Hence $c(f)$ and $c(g)$ belong to the same class of $C2/dC$ 1 in this case too.

Finally, for two arbitrary immersions f and g let us take an immersion h such that both f, h and g, h are pairs of immersions as in the preceding case. Then both $c(f)$ and $c(h)$, as well as both $c(g)$ and $c(h)$, will belong to the same class of $C2/dC$ 1. Hence $c(f)$ and $c(g)$ belong to the same class too.

The theorem is thus proved in all respects.

From the above theorem the following definition is legitimate:

Definition The class in $C2/dC$ 1 of the elements $c(f)$ for any immersion f of G in the plane will be called the *imbedding class* of G and will be denoted by $I(G)$ in what follows.

From the very definition of imbedding it is clear that for G to be imbeddable in the plane it is necessary that
$$I(G) = 0.$$
We shall prove that this condition is not only necessary but also sufficient. For this purpose we shall prove first some preliminary lemmas as follows.

Lemma 1 *Let G' be a subgraph of G. Then the natural restriction will induce morphisms $r1$ and $r2$ so that the diagram below is commutative:*

$$\begin{array}{ccc} C1(G') & \xrightarrow{d} & C2(G') \\ {\scriptstyle r1}\uparrow & & \uparrow{\scriptstyle r2} \\ C1(G) & \xrightarrow{d} & C2(G) \end{array}$$

Moreover, for the morphism thus induced
$$r' : C2(G)/dC1(G) \to C2(G')/dC1(G')$$
we have
$$r'(I(G)) = I(G').$$

The proof is evident and will be omitted.

Lemma 2 *Let G' be a subdivision of G. Then there are natural morphisms $s1$ and $s2$ so that the diagram below is commutative:*

$$\begin{array}{ccc} C1(G) & \xrightarrow{d} & C2(G) \\ {\scriptstyle s1}\uparrow & & \uparrow{\scriptstyle s2} \\ C1(G') & \xrightarrow{d} & C2(G') \end{array}$$

Moreover, for the morphism thus induced
$$s' : C2(G')/dC1(G') \to C2(G)/dC1(G)$$
we have
$$s'(I(G')) = I(G).$$

Proof. Suppose G' is obtained from G by introducing a single new vertex V' on some edge Eq of G with ends Vi and Vj. Define now $s1$ and $s2$ in the following way.

Let us denote the edges V_iV' and V_jV' in G' derived from E_q of G by E_q' and E_q'' respectively. Then for any elements c_1 in $C_1(G')$ and c_2 in $C_2(G')$ we define $s_1 c_1$ in $C_1(G)$ and $s_2 c_2$ in $C_2(G)$ by

$$s_1 c_1 / (V_k, E_r) = c_1/(V_k, E_r), \quad \text{for } E_r \langle \ \rangle E_q,$$
$$s_1 c_1 / (V_k, E_q) = c_1/(V_k, E_q') + c_1/(V_k, E_q''),$$
$$s_2 c_2 / (E_r, E_s) = c_2/(E_r, E_s), \quad \text{for } E_r, E_s \langle \ \rangle E_q,$$
$$s_2 c_2 / (E_q, E_r) = c_2/(E_q', E_r) + c_2/(E_q'', E_r).$$

It is easy to verify that $ds_1 c_1 = s_2 d c_1$ and $s'(I(G')) = I(G)$. The lemma is thus true for this simple case. Since any subdivision of G is formed of a sequence of such elementary subdivisions of the above type, the lemma is proved.

Lemma 3 *For Kuratowski's graphs* $K = K_1$ *or* K_2 *we have*

$$I(K) \langle\ \rangle 0.$$

Proof. Let us consider e. g. the first Kuratowski's graph $K = K_1$. Denote the 5 vertices of K by $V_1, \cdots, V_5$ and immerse K in the plane in the usual way with images W_i of V_i forming a regular pentagon and images of the edges the respective sides and diagonals of the pentagon. The element $c(f)$ of the corresponding immersion is then given by

$$c(f)/(V_1V_3, V_2V_4) = 1,$$
$$c(f)/(V_1V_3, V_2V_5) = 1,$$
$$c(f)/(V_2V_4, V_3V_5) = 1,$$
$$c(f)/(V_1V_4, V_2V_5) = 1,$$
$$c(f)/(V_1V_4, V_3V_5) = 1,$$
$$c(f)/(V_iV_j, V_kV_l) = 0.$$

for any other pair (V_iV_j, V_kV_l) in $D_2(k)$. In other words, we have

$$c(f) = \langle V_1V_3, V_2V_4\rangle + \langle V_1V_3, V_2V_5\rangle + \langle V_2V_4, V_3V_5\rangle$$
$$+ \langle V_1V_4, V_2V_5\rangle + \langle V_1V_4, V_3V_5\rangle.$$

The differential d is defined by

$$d\langle V_1, V_2V_3\rangle = \langle V_1V_4, V_2V_3\rangle + \langle V_1V_5, V_2V_3\rangle, \text{etc.}$$

Consider any element
$$c1 = \text{SUM } Xijk \cdot \langle Vi, VjVk \rangle$$

in $C1(K)$, in which $Xijk = Xikj$ are all mod 2 integers in $Z2$ and the summation is over all triples i, j, k chosen from 1, 2, $\cdots$, 5 which are mutually distinct. If $c(f)$ is the d-image of $c1$, then the following set of equations should be true:

$$X124 + X324 + X213 + X413 = 1,$$
$$X125 + X325 + X213 + X513 = 1,$$
$$X235 + X435 + X324 + X524 = 1,$$
$$X125 + X425 + X214 + X514 = 1,$$
$$X135 + X435 + X314 + X514 = 1.$$
$$X134 + X234 + X312 + X412 = 0,$$
$$X135 + X235 + X312 + X512 = 0, \text{etc.},$$
$$X145 + X245 + X412 + X512 = 0.$$

In all there are 15 such equations. By adding all these equations we get $0 = 1$ since we are working in the domain $Z2$. This proves that $c(f)$ cannot be the d-image of any element in $C1(K)$ or $I(K) \langle \rangle 0$ for $K = K1$. The proof of the case $K = K2$ is similar and will be omitted.

Fundamental Theorem I *For a graph G to be imbcddable in the plane it is necessary and sufficient that*

$$I(G) = 0.$$

Proof. It is enough to prove only its sufficiency. Suppose G is not imbeddable. By Theorem of Kuratowski G should contain a certain subgraph G' which is some subdivision of either $K1$ or $K2$. By Lemmas 2 and 3 we should have $I(G') \langle \rangle 0$. By Lemma 1 we have then a fortiori $I(G) \langle \rangle 0$. Hence $I(G) = 0$ would imply that G is imbeddable. This proves the theorem.

As a complement we have also the following

Theorem 2 *For any element $c2$ in $C2(G)$ belonging to the imbedding class $I(G)$ in $C2(G)/dC1(G)$ there is an immersion f of G such that*

$$c2 = c(f).$$

Proof. This is clearly a direct consequence of the following

Lemma 4 Let c2 be any immersion element of G in the imbedding class $I(G)$. Then for any function $\langle Vk, Eq\rangle$ of $C1(G)$ the element

$$c2' = c2 + d\langle Vk, Eq\rangle$$

is also an immersion element of some immersion of G.

Proof. Let g be any immersion of G defining the element c2. On the interior of the image under g of Eq let us take a small segment L with ends Vi', Vj' which contains no image points of any other edges of G. Replace now L with a simple broken line L' joining Vi', Vj' such that $L + L'$ will form a loop with only image of Vk and with no image of any other vertices in its interior. Let g' be the map of G in the plane which coincides with g everywhere except that L is replaced now by L'. We may also clearly choose L' in such a way that g' is a well-defined immersion. It is now easy to verify that $c(g') = c2'$ and the lemma is thus proved.

Remark All the theorems and proofs here are of an algebraico-topological character, but we have avoided the use of any such terminology as done in the original paper [WU2].It is also clear that the concepts and results in this section may be naturally extended with no essential changes to graphs G not necessarily connected.

3. Reduction of criterion to solvability of linear equations

Consider the graph G and an arbitrary immersion f as before which will determine an immersion element

$$c(f) = \text{SUM } Iqr \cdot \langle Eq, Er\rangle \text{ in } C2, \qquad (1)$$

in which the summation is to be extended over all pairs (Eq, Er) of $D2$ and

$$Iqr = \text{Int}(fEq, fEr). \qquad (2)$$

According to the fundamental theorem in Sect. 2, the planar imbeddability of G depends then on the existence of a function c1 in $C1$ such that

$$dc1 = c(f). \qquad (3)$$

As $C1$ has a basis consisting of functions $\langle Vi, Er\rangle$ with (Vi, Er) running over all pairs in $D1$, we may set such a c1 of $C1$ in the form

$$c1 = \text{SUM } Xir \cdot \langle Vi, Er\rangle, \qquad (4)$$

with summation extended over all pairs (Vi, Er) in $D1$ and Xir unknowns in $Z2$ to be sought for. Form $dc1$ and compare both sides of (3). We then get a system of linear equations

$$(\text{EQN}f) : Xir + Xjr + Xkq + Xlq = Iqr$$

with one equation corresponding to each pair (Eq, Er) in $D2$, the ends of Eq being supposed to be Vi, Vj, and the ends of Er to be Vk, Vl. Fundamental Theorem I in Sect.2 can thus be reformulated in the following form:

Fundamental Theorem II *For a graph G to be imbeddable, it is necessary and sufficient that for an arbitrary immersion f of G in the plane, the system of linear equations $(EQNf)$ possess solutions of the $X's$ in mod 2 integers.*

We remark that the theorem remains true even if G is not connected.

Now the number of equations in $(EQNf)$ is about $Ne \wedge 2$ and that of the unknowns X is about $Nv*Ne$ where Nv and Ne are the numbers of vertices and edges in G respectively. The determination of solvability of this system seems to be thus untractable in appearance. However this is not the case. In fact, owing to the particular form of these equations we can treat the system in a quite feasible algorithmic manner which will lead to a complete solution of both problems P1 and P2.

To see this let us arrange the edges of G in a definite order, say $E1, E2, \cdots, Eq, \cdots, En$, in which $n = Ne$. For each edge Er of G with $r > 1$ let us denote the set of equations in $(EQNf)$ corresponding to pairs (Eq, Er) with $q = 1, 2, \cdots, r-1$ by $(EQNr)$ which may eventually be non-existent. We denote also the subgraph (not necessarily connected) formed of $E1, \cdots, Er$ by Gr. Beginning from $r = 2$, let us arrange the sets of equations $(EQNr)$ successively in an echelon form by the so-called Gaussian elimination with certain sets of equations, to be eventually forsaken. We remark in passing that the met hod of Gaussian elimination occured in fact already in the early Chinese classic *Nine Chapters of Arithmetic* together with introduction of negative numbers which appeared more than 2000 years ago. Now the set $(EQN2)$ may be either empty or consisting of a single equation so that it is already in the echelon form. To start with we shall put $(EQN2')$ to be the same set as $(EQN2)$, empty or not, and introduce a further empty set to be denoted by $(DEL2)$. We set also $G2' = G2$.

Consider now $r > 2$ and put $s = r-1$. Suppose that the sets of equations $(EQNq)$ with $q = 1, 2, \cdots, s$ have already been treated with the result of a set of equations $(EQNs')$ in echelon form as well as a set $(DELs)$ of edges chosen from Gs such that the subgraph Gs' formed by edges in Gs but not in (DELs) is imbeddable in the plane. Remove now from the set, if non-empty, of equations (EQNr) those corresponding to the pairs (Eq, Er) with Eq in (DELs) and denote the set of remaining equations by (EQNr''). If the set (EQNr'') is non-empty, then adjoin this set to (EQNs') and arrange these in echelon form by Gaussian elimination. Two cases may then occur. In the first case the equations newly adjoined will render the whole set a contradictory one. The system of equations is then unsolvable so that the graph Gs' with Er adjoined will become non-imbeddable. We delete thus Er from G and adjoin Er to (DELs) to form (DELr). The subgraph Gr' will be set to be identical to Gs', and the system (EQNr') to (EQNs'). In the second case the reduction to echelon form can be carried out without arriving at contradiction. The system of equations arrived at consisting of (EQNs') and the newly adjoined (EQNr'') in reduced echelon form will then be denoted by (EQNr'). The set (DELr) will remain the same as (DELs) and Gr' will be Gs' adjoined by Er. We remark that as the set (EQNr'') is at most $r-1$ in number and each equation in it has at most 4 unknowns with coefficients in $Z2$, the reduction to echelon form requires actually at most $8*(r-1)$ additions of mod 2 integers.

Finally, if the set (EQNr) or (EQNr'') is empty, then we shall proceed to the next step with (EQNr'), (DELr) the same as (EQNs'), (DELs), and Gr' as Gs' with Er adjoined.

From the above we get thus the following

Theorem 3 *There is an algorithmic procedure which permits to determine in a finite number of steps whether a given graph G is imbeddable or not, and in the case it is not imbeddable, a set of edges should be deleted from the given graph so that the remaining graph is imbeddable.*

The above method settles thus both the problems $P1$ and $P2$ and can be easily programmed. To apply it we have to choose first an arbitrary immersion of the graph G, form successively the set of equarions (EQNr) and proceed as indicated above, As already remarked, the whole procedure requires at most

$$Na = \text{SUM } 4*(Ne-1)*Ne < 4*Ne \wedge 2$$

mod 2 additions and is thus quite feasible. The only defect is that a large amount of memory space may be required. We shall discuss this matter in later sections.

4. An alternative reduction of criterion to solvability of linear equations

In the original paper ⟨WU1⟩ the author has described a method of reducing the criterion of imbeddability to the solvability of a system of linear equations on $Z2$, which is a little different from that given in Sect.3. Though the proof of this reduction is rather involved, it has however the advantage of being able to greatly reduce the number of unknowns in the equations. What is more important is that this method will lead to a complete solution of problems P1–4, in comparison with the one in Sect. 3 which permits to solve only problems: P1–2. We repeat the remark already made in the introduction that it is problem P3 that is the decisive part in view of applications.

In order to explain this method we shall first introduce some notions as well as notations. Henceforth G will be supposed to be connected.

By a *tree* of the graph G supposed connected we shall mean a maximal one belonging to G, i.e. one passing through all vertices of G. Let a tree T be taken and fixed in what follows.

With respect to tree T of G the vertices will be divided into two classes: *internal* ones and *terminal* ones. The edges of G will also be divided into two classes: those belonging to the tree and those not. We shall call these *tree-edges* and *external edges* and denote them by $Eu, Ev, Ew, \cdots$ and $Ea, Eb, Ec, \cdots$ respectively.

Among the terminal vertices of T we shall choose one as the root of T which will be denoted by O henceforth.

Without loss of generality we shall make the assumption that no external edges issue from O. In fact, in the contrary case we may adjoin an extra edge to the graph G with one end at O and the other free. We may then take that free end as the new root of the new graph. The problems are actually the same for the new graph and the original one so far as imbeddability is concerned. Hence we shall suppose that the above device has been adopted in what follows so that the above assumption is always verified.

For any vertex Vi of G different from O there is a unique path belonging to T which leads from Vi to O and will be denoted by Pi. For any external edge Ea there is also a unique path belonging to T which joins the two ends of Ea and will be denoted by Pa. The cycle of G formed by Ea and Pa will then be denoted by Ca.

For any two vertices Vi, Vj ($\langle \rangle O$) the two paths Pi, Pj will begin to meet first at some vertex in running toward O which will then be called the *V-meet* of Vi, Vj or of Pi, Pj. For any tree-edge Eu with ends Vi, Vj, one of them, say Vi, will have the path Pi containing the other end Vj. We shall then call Vi the *head* and Vj and the *tail* of Eu.

Each tree-edge Eu will divide the tree T into two disconnected parts, say $T'u$ and $T''u$. One, say $T'u$, will contain the head of Eu and the subtree formed by $T'u$ and Eu will be denoted by Tu. The set consisting of Eu as well as all edges with two ends one in $T'u$ and the other in $T''u$ will be denoted by CSu. In the network theory a set of edges is called a *cut-set* of G if in removing it G will split into two or more disconnected parts. The set CSu is such a cut-set and in the network theory it is proved that the collection of sets CSu corresponding to all the tree-edges Eu form a basis of all cut-sets in an evident sense, cf. e.g. [SB1].

For any two sets of edges $S1$, $S2$ of G the function in $C2$ taking the value 1 on all pairs (Eq, Er) of $D2$ with Eq in $S1$ and Er in $S2$ and the value 0 otherwise will be denoted by $[S1, S2]$. In other words

$$\langle S1, S2 \rangle = \text{SUM}\langle Eq, Er \rangle,$$

in which the summation is to be extended over all pairs (Eq, Er) as above. The following lemma is now readily proved (cf. [WU 1]) :

Lemma 5 *The subgroup $d\,C\,1$ of $C\,2$ has a set of generators (not necessarily a basis) consisting of elements*

$$\langle CSu, CSv \rangle \ and \ \langle CSu, Ea \rangle,$$

corresponding to all pairs (Eu, Ev) and (Eu, Ea), disjoint or not, respectively.

Introduce now sets of variables or unknowns on $Z2$ as follows. To each unordered pair (Eu, Ev) of tree-edges disjoint or not is associated an unknown $Xuv\ (=Xvu)$.

To each pair (Eu, Ea) of a tree-edge Eu and an external edge Ea disjoint or not is associated an unknown Yua

By the lemma above and Fundamental Theorem II in Sect. 2 it follows that for G to be imbeddable it is necessary and sufficient that for an arbitrary T-immersion f of G the following system of linear equations in $Z2$ be solvable in the unknowns X and Y:

$$\text{SUM } Xuv \cdot \langle CSu, CSv \rangle + \text{SUM } Yua \cdot \langle CSu, Ea \rangle = \text{SUM } Iqr \cdot \langle Eq, Er \rangle,$$

in which the various summations are to be extended over respective ranges. Compare the terms of both sides and note that by the very definition of a T-immersion $Iqr = 0$ when Eq or Er or both are tree edges, we get:

$$Xuv = 0, \text{for } (Eu, Ev) \text{ in } D2,$$
$$Yua = \text{SUM1 } Xuv, \text{for } (Eu, Ea) \text{ in } D2,$$
$$Iab = \text{SUM2 } Yua + \text{SUM3 } Yub + \text{SUM4 } Xuv.$$

The various summations are respectively extended over the ranges as follows:

SUM1 over Ev in Pa,

SUM2 over Eu in Pb,

SUM3 over Eu in Pa,

SUM4 over pairs (Eu, Ev), disjoint or not, with Eu in Pa and Ev in Pb.

Corresponding to each pair (Eu, Ea) of a tree-edge Eu and an external edge Ea with Eu, Ea disjoint or not let us introduce a new unknown Xua in $Z2$ by setting

$$Xua = \text{SUM1 } Xuv + Yua,$$

so that by equations about Yua above,

$$Xua = 0, \text{for } (Eu, Ea) \text{ in } D2.$$

The equation about Iab will then become

$$Iab = \text{SUM2}(\text{SUM1 } Xuv + Xua) + \text{SUM3}(\text{SUM5 } Xuv + Xub) + \text{SUM4 } Xuv,$$

with SUM5 given by

SUM5 *over Ev in Pb*.

As the terms SUM2 SUM1 Xuv and SUM3 SUM5 Xuv are actually the same they cancel each other in $Z2$. Taking into account the equation $Xua = 0$ for (Eu, Ea) in $D2$, we get then

$$\text{SUM0 } Xuv + \text{SUM}' \ Xua + \text{SUM}'' \ Xub = Iab, \tag{If}$$

in which the various summations are to be extended over ranges as follows:

SUM0 over pairs (Eu, Ev) non-disjoint with Eu in Pa and Ev in Pb,

SUM$'$ over pairs Eu, Ea with Eu in Pb and (Eu, Ea) non-disjoint,

SUM$''$ over pairs Eu, Eb with Eu in Pa and (Eu, Eb) non-disjoint.

This leads to the following

Theorem 4. *For a graph G to be imbeddable it is necessary and sufficient that for an arbitrary tree T and an arbitrary T-immersion f of G the system of equations* (If) *corresponding to pairs (Ea, Eb) in D2 be solvable in the unknowns X in Z2.*

摘　　要

本文是 1967 年以来作者只用中文发表的所得结果的一个英文综述, 在文中证明了连通线性图可嵌入平面的一个充要条件是某一组 mod 2 系数的线性方程组有解. 在该方程组有解因而线性图可嵌入平面时, 又可考虑另一组仍为 mod 2 系数的二次方程组, 并根据这两组方程必然存在的共同解答来作出图的具体嵌入. 若图的顶点数与棱数各为 N_v 与 N_e, 而顶点的最大次数为 m, 则这些方程中的未知数个数最多为

$$(m-3)*N_e + N_v,$$

且在决定能否嵌入时只须用到不超过 $4*N_e \wedge 2$ 的 mod 2 加法即可. 因之这一方法容易编成程序且是切实可行的.

On the Planar Imbedding of Linear Graphs (Continued)*

5. Further reduction of fundamental system of linear equations

The fundamental system of linear equations (If) in preceding Sect 4 can actually be put in a much simpler form. For this purpose let us denote by NT a certain neighborhood of T in G sufficiently small so that the T-immersions considered will be some imbeddings when restricted on NT and that all possible intersections of images of disjoint external edges are not in NT. We shall call these T-immersions also *NT-immersions*.

Lemma 6 *For NT-immersions of G the immersion elements are already completely determined by the restricted imbeddings of the neighborhood NT of the respective immersions.*

Proof. For any external edge Ea with ends Vi, Vj let us take two points on Ea lying in the neighborhood NT and denote them by Via, Vja. These points will be taken so near to the tree that the edge Ea will split into three parts from Vi to Via, from Via to Vja, and from Vja to Vj, disjoint from each other except at possible common ends. These parts will be denoted by Eia, Ea', Eja respectively.

Consider now any two NT-immersions. f and g which coincide on NT as imbeddings. For any pair of edges (Ea, Eb) in $D2$ with Vi, Vj the ends of Ea and Vk, Vl the ends of Eb let us denote the intersection number $\text{Int}(fEa, fEb)$ still by Iab while $\text{Int}(gEa, gEb)$ by Jab. We have then

$$Iab = \text{Int}(fEa, fEb') = \text{Ord}(fVkb, fCa) + \text{Ord}(fVlb, fCa),$$
$$Jab = \text{Int}(gEa, gEb') = \text{Ord}(gVkb, gCa) + \text{Ord}(gVlb, gCa).$$

As f and g coincide on NT, so on $Z2$,

$$fCa + gCa = fEa' + gEa', \quad \text{say} = Ca',$$

---
*本文原载《系统科学与数学》, 1986, 6(1): 23-35.

and we have therefore

$$Iab + Jab = \mathrm{Ord}(fVkb, Ca') + \mathrm{Ord}(fVlb, Ca').$$

It follows that

$$Iab = Jab,$$

since $fVkb$ and $fVlb$ are the ends of the broken line $fEkb + fPb + fElb$ disjoint from the polygon Ca'. The immersion elements $c(f)$ and $c(g)$ of f and g are therefore the same. The proof is completed.

In order to avoid tedious verifications in the case that an external edge ends at some internal vertex of the tree, we shall adopt the following devices: If the external edge Ea has some internal end(s) of the tree as an end Vi (or both ends Vi and Vj), then we shall replace, with notations as in the proof of the above lemma, the edge Ea with two (or three) edges $VaVia$, and $ViaVj$ (or $VaVia$, $ViaVja$, and $VjaVj$). By the lemma above as well as Lemma 2 of Sect.2 we see that it is immaterial to replace the original graph by this new one. We shall suppose in what follows that such a modification of the graph has already been done so that we may assume the following conventions for the graph G have been observed :

Convention 1. The root O of the tree T is a terminal vertex.

Convention 2. The ends of any external edge are both terminal vertices.

Let us call an unordered pair of tree-edges (Eu, Ev) a *redundant* one if the tail of say Eu is the head of Ev. The corresponding unknown Xuv is then also said to be *redundant*. Consider now any set of mod 2 integers $(Aqrs)$ corresponding to all unordered triples of tree-edges (Eq, Er, Es) with one end in common for which each $Aqrs$ is independent of the order of indices (q, r, s). Let us also put for each such triple

$$Xqrs = Xqr + Xqs + Xrs,$$

which is also independent of the order of the indices (q, r, s). We have then

Lemma 7 *If the set of equations*

$$Xqrs = Aqrs$$

corresponding to all unordered triples of edges (Eq, Er, Es) with a common end is solvable for $Xuv = Xvu$ in $Z2$, then the same set of equations is also solvable with all redundant unknowns $Xuv = 0$.

Proof. Let $(Xuv) = (Auv)$ with $(Auv) = (Avu)$ corresponding to each unordered pair of edges (Eu, Ev) with common end be a solution of the above system of equations. For each such pair (Eu, Ev) with common end Vm let Eq be the edge on the path Pm with Vm as head. Then we see easily that

$$Xuv = 0 \text{ for } (Eu, Ev) \text{ redundant and}$$
$$Xuv = Auq + Avq + Auv \text{ otherwise}$$

is also a solution of the above system of equations and the lemma is thus proved.

We are now in a position to simplify the system of equations (If) of Theorem 4 in Sect.4. We shall denote the equation in (If) corresponding to the pair (Ea, Eb) in $D2$ by (EQab) and the ends of Ea, Eb by Vi, Vj and Vk, Vl respectively.

First let us remark that owing to conventions 1 and 2 the terms in SUM' Xua and SUM" Xub in the equation (EQab) are no more existent. For the terms in SUM0 Xuv we distinguish three cases.

Case 1. Pa, Pb are disjoint.

We see that SUM0 is nonexistent and fEa, fEb do not meet. So the equation (EQab) becomes $0 = 0$ and is redundant.

Case 2. Pa, Pb meet at a single vertex Vm.

Let the tree-edges with end Vm on the paths from Vi, Vj, Vk, Vl to Vm be respectively Ep, Eq, Er, and Es. Then we see that SUM0 reduces to four terms

$$Xpr + Xps + Xqr + Xqs.$$

The equation (EQab) can thus be written in the form

$$Xpqr + Xpqs = Iab.$$

Case 3. Pa, Pb have a tree-path in common with end vertices Vm, Vn.

We may suppose that Vm is on the tree-path $ViVk$ and Vn on $VjVl$. Let the edges with end Vm on the tree-paths $ViVm$, $VkVm$ and $VmVn$ be respectively Ep, Er and Ew. Similarly, let the edges with end Vn on the tree-paths $VjVn$, $VlVn$ and $VnVm$ be respectively Eq, Es and Ez. Then we see that the equation (EQab) may be written in the form

$$Xprw + Xqsz = Iab.$$

Suppose now G is imbeddable so that (If) is solvable in the unknowns $Xuv = Xvu$. In view of the above analysis of the form of the equations (EQab) in (If) it follows from Lemma 7 that the system (If) will also be solvable in unknowns $Xuv = Xvu$ with all redundant ones $= 0$. Moreover, for any different terminal vertices Vi, Vj, both different from root O with Vm the V-meet of the paths Pi and Pj, different from O owing to our conventions, let Er, Es be the edges on Pi, Pj with tail Vm and set

$$Xij = Xrs(= Xji).$$

Then, with all redundant $Xuv = 0$ it is easily verified that the left-hand side of the equations (EQab) either in Case 2 or in Case 3 can always be written in the form

$$Xik + Xjk + Xil + Xjl$$

which may eventually be reduced to only two terms.

It follows that the system of equations (If) may be replaced by a system (Xf) below which is much simpler in form and the Fundamental Theorem II may also be re-stated as :

Fundamental Theorem II' *A graph G is imbeddable if and only if, given a tree T, a root O, and a T-immersion f, the system of equations*

$$Xik + Xil + Xjk + Xjl = Iab, \qquad \text{(Xf)}$$

corresponding to pairs (Ea, Eb) *in* $D2$ *with* Vi, Vj *ends of* Ea *and* Vk, Vl *ends of* Eb, *is solvable in* $Z2$.

We remark that, under the conventions above, each Xij occuring in equations (Xf) is some Xrs for a pair of non-disjoint tree-edges (Er, Es) having a common tail. We shall call all such pairs (Er, Es) *admissible pairs* in what follows.

We see that each equation of (Xf) involves actually at most 2 or 4 unknowns of X and eventually has the trivial form $0 = 0$. Morever, the number of unknowns of X are readily estimated as in the following

Theorem 5 *If the maximum order of vertices in the graph G is m, then the number of unknowns of X occurring in the fundamental systemof equations (Xf) is at most*

$$Nx = (m - 3) * Ne + Nv,$$

in which Nv and Ne are respectively the original number of vertices and edges of G before modification.

Proof. Let Ok be the number of vertices of order k in G. Then we have

$$\text{SUM } k * Ok = 2 * Ne,$$

the summation being over $K >= 1$. Now to each vertex of order k the associated number of unknowns of X that may occur in the equations (Xf) is clearly at most $(k-1)*(k-2)/2$. Hence we have

$$\begin{aligned}
Nx <&= \text{SUM}((k-1)*(k-2)/2)*Ok \\
&= \text{SUM}k*(k-2)*Ok/2 - 1/2*(\text{SUM}(k-2)*Ok) \\
<&= (m-2)*Ne - Ne + \text{SUM } Ok \\
&= (m-3)*Ne + Nv.
\end{aligned}$$

That Nv, Ne may be taken to be the original numbers of the unmodified graph is also clear.

To determine the imbeddability of G we can now proceed as in Sect. 3 with the result of getting a set of edges that is to be deleted and a set of solutions of remaining equations in mod 2 integers of the unknowns X. The operations require only mod 2 additions at most $4*Ne \, \Lambda \, 2$ in number as before. The number of unknowns has however been reduced so that much les memory space will be required. In particular, if the graph G has only vertices of order $<= 3$, then the number of unknowns will be $<= Nv$, the number of vertices of G.

The great advantage of this method lies in reality in the fact that actual imbedding of an imbeddable graph G, or more generally the imbedding of the imbeddable graph which remains after removal of a certain set of edges, can be constructed from the set of solutions obtained from the fundamental system of equations. This will form the object of study in the next sections.

6. Geometrical interpretation of unkowns X and rotation numbers associated to a T-immersion

As stated at the end of last section, the solutions of the fundamental system of equations (Xf) for a graph G supposed to be imbeddable or become imbeddable after removal of certain edges will lead to a method of actual imbedding of such a graph. To see this we shall first give in this section some geometrical interpretation of the unknowns X involved in these equations. In fact, by Lemma 6 of Sect. 5 the nature of the T-immersion f will actually be determined by f on NT and this in turn will be determined by how the edges at a common end are mutually situated when immersed by f. This suggests thus the introduction of the following notions.

Let $L1$, $L2$, $L3$ be three simple broken lines in the plane disjoint from each other except that they have one end in common. We shall attach then to this ordered triple of lines a *rotation number* (in $Z2$).

$$\mathcal{R}(L1, L2, L3) = 0 \text{ or } 1$$

according as in passing from $L1$ to $L3$ through $L2$ we have to turn around their common end in a counter-clockwise or in a clockwise sense.

Consider now an NT-immersion f of G with conventions 1, 2 observed. For any admissible pair of edges (Er, Es) with common tail Vm we set then by definition

$$Rrs(f) = R(fPm, fEr, fEs).$$

For each pair of vertices Vi, Vj different from O which lead to the admissible pair (Er, Es) with Er on Pi and Es on Pj having a common tail we shall set by definition

$$Rij(f) = Rrs(f).$$

Remark that the order of the indices are important in that

$$Rrs(f) = Rsr(f) + 1, \quad Rij(f) = Rji(f) + 1.$$

Theorem 6 *Let f be a T-immersion of G with conventions 1, 2 observed. Then for any pair of edges (Ea, Eb) in $D2$ with Vi, Vj ends of Ea and Vk, Vl ends of Eb, we have*

$$Iab(f) = Rik(f) + Rjk(f) + Ril(f) + Rjl(f).$$

Remark *The equation is of the same form as the corresponding one in the fundamental system of equations (Xf). However the numbers $Rik(f)$ do not form a solution of the system (Xf) since $Rik(f) <> Rki(f)$ while we are seeking for solutions with $Xik = Xki$.*

Proof. Let Vm be the V-meet of the paths Pi, Pj. Consider first the vertex Vk. According as fO and fVk are interior or exterior to the cycle fCa, and according as the path Pk does not meet Ca or first meets Ca on Pi or Pj, there are in all 12 cases to consider. We verify easily that in all cases we shall have

$$Rik(f) + Rjk(f) = \text{Ord}(fVk, fCa) + \text{Ord}(fO, fCa).$$

Similarly we have

$$Ril(f) + Rjl(f) = \text{Ord}(fVl, fCa) + \text{Ord}(fO, fCa).$$

Hence we get

$$Rik(f) + Rjk(f) + Ril(f) + Rjl(f)$$
$$= \text{Ord}(fVk, fCa) + \text{Ord}(fVl, fCa)$$
$$= Iab.$$

Consider now two NT-immersions f and g of G. For any admissible pair of edges (Er, Es) let us set by definition

$$Wrs(f,g) = Rrs(f) + Rrs(g).$$

For any pair of vertices Vi, Vj different from O and leading to the admissible pair (Er, Es) we set then by definition

$$Wij(f,g) = Rij(f) + Rij(g),$$

or, what is the same,

$$Wij(f,g) = Wrs(f,g).$$

Remark that unlike the $R's$ the numbers W are no more dependent on the order of the indices

$$Wrs(f,g) = Wsr(f,g), \quad Wij(f,g) = Wji(f,g).$$

If the common tail of the admissible pair Er, Es is at Vm, then

$$Wrs(f,g) = 0 \text{ or } 1$$

according as the configurations (fPm, fEr, fEs) and (gPm, gEr, gEs) have the same sense of rotations or not. Hence the set of numbers $Wrs(f, g) = Wsr(f, g)$ corresponding to all admissible pairs (Er, Es) serves to compare the configurations of the two imbeddings f/NT and g/NT. More precisely we have the following

Theorem 7 *Let f, g be two NT-immersions of G. Suppose the fundamental system of equations (Xg) corresponding to g is solvable and has a solution*

$$(Xrs) = (Brs),$$

in which $Brs = Bsr$ are numbers in $Z2$ corresponding to all admissible pairs of edges (Er, Es). Then the fundamental system of equations (Xf) corresponding to f is also solvable and has a solution

$$(Xrs) = (Ars),$$

in which
$$Ars = Brs + Wrs,$$
where we have put for simplicity
$$Wrs = Wrs(f, g).$$

Proof. Let us set by definition
$$Ars = Brs + Wrs(= Asr).$$
Set also by definition
$$Bij = Br, \quad Aij = Ars,$$
if the pair of vertices Vi, Vj different from O will lead to the admissible pair of edges(Er, Es). Then we shall have also
$$Aij = Bij + Wij(= Aji).$$

Consider now any pair of edges (Ea, Eb) in $D2$ with Vi, Vj ends of Ea and Vk, Vl ends of Eb. Write Iab, Jab for their respective intersection index under f and g as before. As (Brs) is a solution of the system (Xg), so we have
$$Jab = Bik + Bjk + Bil + Bjl.$$
By Theorem 6 we have also
$$Jab = Rik(g) + Rjk(g) + Ril(g) + Rjl(g),$$
$$Iab = Rik(f) + Rjk(f) + Ril(f) + Rjl(f).$$

Adding all these three equations together and taking into account the definition of A and W, we get
$$Iab = Aik + Ajk + Ail + Ajl.$$

This shows that the set of numbers (Ars) in $Z2$ with $Ars = Asr$ forms a solution of (Xf) and the theorem is proved.

Suppose now in particular that G is imbeddable with g an imbedding not only of NT but also of G as a whole. Then we have $Jab = 0$ for all pairs in $D2$ so that $Xrs = 0$ for all admissible pairs (Er, Es) forms a trivial solution of the system (Xg). By the above theorem
$$(Xrs) = (Wrs)$$

will form then a solution of the system (Xf). From the meaning of Wrs we have therefore the following

Geometrical Interpretation of the unknowns Xrs:

The set $Xrs = Xsr$ will serve as a set of indicators whether for each admissible pair of edges (Er, Es) with common tail Vm their images under f should be modified to change the sense of rotations of the triple (fPm, fEr, fEs) so that the modified immersion of T may be extendable to an actual imbedding of the whole graph G. See however the next section.

7. Quadratic relations among the unknowns X

By Fundamental Theorem II we know that for an arbitrary NT-immersion f of G, if the fundamental system of equations (Xf) possesses a solution

$$(Xrs) = (Ars),$$

with $Ars = Asr$ corresponding to all admissible pairs of edges (Er, Es) in $D2$, then G is imbeddable. From the preceding section it seems further that from this solution we would get an actual imbedding of G by modifying the NT-immersion f to another one g in changing the mutual rotational relationships of edges at the same vertices according to the formulae

$$Wrs(f, g) = Ars.$$

However, this is entirely not the case. In fact, though G is imbeddable if the system of equations (Xf) is solvable, not every solution of (Xf) will lead to an actual imbedding of G in the above manner. The reason may be seen as follows.

Let us consider any triple of edges Er, Es, Et with same tail Vm. For any immersion g we have then a set of 3 rotation numbers in $Z2$, *viz.*

(Rg) $\qquad\qquad Rrs(g), Rst(g), Rtr(g)$.

As each R may take a value of 0 or 1, so apparently there would be 8 such sets of values to be taken for (Rg). However there are only 6 different types of orientational relationships of the edges Er, Es, Et and Pm under g. This shows that among the 8 sets of values of (Rg) only 6 will actually be geometrically realizable. In fact, (Rg) can never take up the sets of values (0, 0, 0) and (1, 1, 1). The problem thus arising is to pick out these 6 sets of values among the 8 sets. The solution of this problem will be furnished by the following device introduced in the original paper [WU4].

For an immersion g and 3 edges Er, Es, Et with same tail Vm let us set by definition

$$Qrst(g) = Rrs(g) * Rrt(g) + Rst(g) * Rsr(g) + Rtr(g) * Rts(g).$$

Remark that though the numbers $Rrs(g)$, etc. depend on the order of indices (r, s) etc., $Qrst(g)$ is independent of the order of the indices (r, s, t). We have now the following

Lemma 8 *The rotation numbers (Rg) satisfy always the relation*

$$Qrst(g) = 1.$$

Proof. Let us first remark that if g is such that in turning around the common end gVm, we shall get successively gPm, gEr, gEs, gEt in the counter-clockwise order, then we have $Rrs(g) = 0, Rst(g) = 0, Rtr(g) = 1$ so that $Qrst(g) = 1$.

Suppose next that $Qrst(g) = 1$ for a certain configuration of gPm, gEr, gEs, gEt in the plane with e.g. gEr, gEs neighboring to each other in the arrangement. Let us interchange the orientational relationship between gEr, gEs but leave the others unchanged to get a new immersion g'. Then we have

$$Rrs(g') = Rrs(g) + 1,$$
$$Rst(g') = Rst(g), \quad Rtr(g') = Rtr(g).$$

It follows that

$$Qrst(g') = Qrst(g) + Rrt(g) + Rst(g) = Qrst(g),$$

since, with gEr, gEs neighboring to each other in the plane, $Rrt(g) = Rst(g)$. This proves the lemma since any other configuration of Pm, Er, Es, Et under any immersion may be got from the one under g by a number of such interchanges of immersed neighoring edges.

For any immersion f of G let us now introduce by definition a system of quadratic forms

$$\begin{aligned}Qrst(f, X) =& (Xrs + Rrs(f)) * (Xrt + Rrt(f)) \\ &+ (Xst + Rst(f)) * (Xsr + Rsr(f)) \\ &+ (Xtr + Rtr(f)) * (Xts + Rts(f))\end{aligned}$$

corresponding to each triple (Er, Es, Et) with a common tail. Introduce also the system of quadratic equations

$$Qrst(f,X) = 1 \qquad (\text{Qf})$$

corresponding to all such triples. Note that in the equations $Xrs = Xsr$ while $Rrs(f) = Rsr(f) + 1$, etc. However $Qrst(f, x)$ is independent of the order of indices (r, s, t). We have now the following

Fundamental Theorem III *If corresponding to a T-immersion of G the fundamental system of equations* (Xf) *is solvable, then the systems of equations* (Xf) *and* (Qf) *taken together are also solvable.*

Proof. As the system (Xf) is solvable, by Fundamental Theorem II′ G is imbeddable with a certain g as an imbedding of G as a whole. Corresponding to each admissible pair of edges (Er, Es) let us put

$$Wrs = Wrs(f, g)$$

for simplicity. Then as in the proof of Theorem 7 of the last section, the system (Xf) will have a solution

$$(Xrs) = (Wrs).$$

By Lemma 8 above the set of numbers $(Rrs(g))$ will satisfy the relations

$$Qrst(g) = 1$$

corresponding to all triples of edges (Er, Es, Et) with common tails. As

$$Wrs = Rrs(f) + Rrs(g),$$

we see that $Qrst(f, X)$ will become $Qrst(g)$ when Xrs, etc. are substituted by Wrs, etc. This shows that $(Xrs)=(Wrs)$ will satisfy both systems of (Xf) and (Qf).

8. Actual imbedding of imbeddable graphs

We are now ready to settle problem P3 of actually imbedding an imbeddable graph G in the plane, assuming that certain edges have already been removed to make G the remaining imbeddable part if necessary. For this purpose we shall first prove a converse of Lemma 8 of the preceding section, *viz.*

Lemma 9 Let (Nrs) be a set of numbers in $Z2$ with $Nrs = Nsr + 1$ corresponding to all admissible pairs of edges (Er, Es) with same tail which satisfies the relations

$$Nrs * Nrt + Nst * Nsr + Ntr * Nts = 1 \qquad \text{(Nrst)}$$

corresponding to all triples of edges (Er, Es, Et) with common tails. Then there is an immersion g of G such that the rotation numbers under g coincide with the corresponding numbers N, i.e.

$$Rrs(g) = Nrs$$

for all admissible pairs of edges (Er, Es) of G.

Proof. Let us consider the simple case that G consists of an edge OVm and a finite set of edges

$$Er, Es, \cdots, Et, \cdots, \qquad \text{(E)}$$

all having Vm as common end. The tree T is then the same as G and O will be chosen as the root. If the number n of the edges in (E) is $n = 3$, then it is clear by the preceding section that such immersion (in fact an imbedding) g of $G = T$ in the plane exists. We shall now proceed to prove this in general by induction on n.

Suppose thus the number of edges in (E) is $n > 3$. By induction there is an immersion g' of $G = T$ such that for all pairs of edges (Ep, Eq) chosen from the set

$$Es, \cdots, Et, \cdots \qquad \text{(E')}$$

we have

$$Rpq(g') = Npq.$$

Suppose that in turning around the common end $g'Vm$ in a counter-clockwise sense on starting from $g'Pm$ ($Pm = OVm$) we shall pass in succession

$$\cdots, g'Es, \cdots, g'Ep, g'Eq, \cdots, g'Et, \cdots.$$

Suppose that among the numbers

$$\cdots, Nsr, \cdots, Npr, Nqr, \cdots, Ntr, \cdots$$

in this order the first non-zero number is Nqr so that

$$\cdots = Nsr = \cdots = Npr = 0, \quad \text{while } Nqr <> 0.$$

The equation in N corresponding to a triple of indices (r, q, t) with Et in the partial set of edges after Eq in the above order is given by

$$Nrq * Nrt + Nqt * Nqr + Ntr * Ntq = 1.$$

As

$$Nqt = Rqt(g') = 0, \quad Nrq = Nqr + 1 = 0, \quad Ntq = Rtq(g') = 1$$

we get

$$Ntr = 1, \quad \text{or } Nrt = 0.$$

Modify now the immersion g' to an immersion g such that g will be the same as g' on Pm and on all edges in (E') while gEr will be brought to a position between $gEp = g'Ep$ and $gEq = g'Eq$. Then we see that for any edge Es in the partial set of edges before Ep in the above order and any edge Et in the partial set of edges after Eq in that order,

$$Rrs(g) = 1 = Nrs, \quad Rrt(g) = 0 = Nrt,$$
$$Rrp(g) = 1 = Nrp, \quad Rrq(g) = 0 = Nrq.$$

For the other number $R's$ we have say $Rst(g) = Rst(g') = Nst$. Hence g will have its rotation numbers all equal to the corresponding numbers N. The induction is thus completed and the lemma is proved for the above special graph G.

For the general graph we shall proceed in just the same manner with the modification that each time we bring a certain edge Er to a new position, we shall bring the whole sub-tree Tr to such new position at the same time. Arrange now the vertices different from O in a definite order and treat each vertex in turn as for the special graph above, with the above modification taken into due account. The rotation numbers of the new immersion for admissible pairs of edges with common tail at that vertex will be identical with the corresponding numbers N. Remark that the interchanges at one vertex will not affect the results of interchanges at other vertices. Hence in proceeding successively we shall finally arrive at a T-immersion with the desired property. The lemma is thus completely proved.

We have now the following

Fundamental Theorem IV *If corresponding to a T-immersion f of G the fundamental systems of equation* (Xf) *and* (Qf) *taken together possess in $Z2$ a solution*

$$(Xrs) = (Ars)$$

with $Ars = Asr$ corresponding to all admissible pairs of edges (Er, Es) with common tails, then there is an imbedding g of G as a whole in the plane with

$$Rrs(g) = Ars + Rrs(f)$$

for all such pairs (Er, Es).

Proof. Set for each admissible pair of edges (Er, Es)

$$Nrs = Ars + Rrs(f) \ (= Nsr + 1).$$

Then by the hypothesis of the theorem the set of numbers Nrs will satisfy all relations of the form (Nrst) corresponding to triple of edges (Er, Es, Et) with common tails. By Lemma 9 above there will be some T-immersion g of G with

$$Rrs(g) = Nrs$$

for all admissible pairs (Er, Es).

By Theorem 7 of Sect. 6, the fundamental system of equations (Xg) corresponding to g will have now a solution given by

$$Xrs = Ars + Wrs,$$

where

$$Wrs = Wrs(f,g) = Rrs(f) + Rrs(g).$$

Consequently (Xg) will have a solution identical to 0:

$$Xrs = 0$$

for all admissible pairs of edges (Er, Es). It follows from Fundamental Theorem II' in Sect. 5 that for any pair of external edges (Ea, Eb) in $D2$ we should have

$$Iab = 0$$

or

$$\mathrm{Int}(gEa, gEb) = 0.$$

Arrange now all the external edges of G in a definite order, say

$$Ea, \cdots, Eb, \cdots, Ec, Ed, \cdots . \tag{E}$$

Our aim is to extend the part of the T-immersion g restricted to T successively to the external edges of (E) to get each time an imbedding of T with successively adjoined edges as a whole. The final imbedding achieved in this way will then be a required imbedding of G in the plane as a whole.

Such an extension to Ea is trivial. Suppose that

$$Ea, \cdots, Eb, \cdots, Ec \qquad (E')$$

in the ordered set (E) have been extended so that we have an imbedding g' of $G' = T+$ (E') in the plane with g'/T identical to g. Let us try to extend g' to an imbedding including the next new edge Ed in the set (E). Consider any external edge disjoint from Ed in (E''), say Eb. By Lemma 6 of Sect. 5, we have for the pair (Eb, Ed) in $D2$

$$\text{Int}(g'Eb, g'Ed) = \text{Int}(gEb, gEd) = 0.$$

This means that if the ends of Ed are Vi and Vj, then we should have

$$\text{Ord}(g'Vi, g'Cb) = \text{Ord}(g'Vj, g'Cb).$$

Consequently $g'Vi$ and $g'Vj$ will lie in the same region in the plane separated by $g'T$, $g'Ea, \cdots, g'Ec$ of $g'G'$. We may thus join $g'Vi$ and $g'Vj$ by a simple broken line not meeting $g'G'$ except at the two ends. We extend than g'' to Ed by taking this broken line to be the image $g''Ed$. This achieves the induction and proves the thorem.

9. Procedure of solving problems P1—3 for a graph

From the developments of the last sections it is now clear how to solve problems P1—3 for a given graph G. The procedure will be as follows.

Step 1 *Choose an arbitrary tree T of G as well as a root O. Modifications may be made according to Sect. 5 if required.*

Step 2 *Take an arbitrary T-immersion f of G.*

Step. 3 *Form the fundamental system of equations (Xf) successively and solve in the way as shown in Sect. 5. We get then a set DEL of edges to be removed from G to render the remaining graph G' imbeddable. Denote the restriction of f to G' by f'. As no ambiguity can occur we shall denote G' and f' again by G and f. The set of solutions of corresponding fundamental equations (Xf) will be denoted by (S).*

Step 4 *Form the system of quadratic equations (Qf) for G (i.e.(Qf') for G') and verify whether each solution in the set (S) is also a solution of (Qf) or not. By Fundamental Theorem III of Sect. G, such solutions necessarily exist.*

Step 5 *For any solution of* (Xf) *and* (Qf) *taken together, modify f/T to a T-immersion g of G as in Sect. 8. Such a T-immersion g may then be extended to get an imbedding of G in the plane as a whole as shown in the Fundamental Theorem IV of Sect. 8.*

Remark. By introducing new unknowns and new system of equations it can be shown that the totality of all possible imbeddings of the imbeddable graph essentially different from each other will be obtained in correspondence with the solutions of the three systems of equations taken altogether. This gives the solution of problem P4 as stated in Sect. 1. We shall not however enter into this and will leave the details to the original paper [WU4].

References

[AP1]   Auslander, L. & Parter, V.. On imbedding graphs in the sphere. *J. Math. Mech.*, 1961, 10: 517-523.

[HT1]   Hopcroft, J. & Tarjan, R.. Efficient planarity testing. *JACM*, 1974, 21: 549-568.

[KU1]   Kuratowski, C.. Sur Ie problème des courbes gauches en topologie. *Fund. Math.*, 1930, 15: 271-283.

[L1]   Liu Yan-pei. Modulo-2 programming and planar imbedding. *Acta Math. Appl. Sinica*, 1978, 1: 321-329 (in Chinese).

[L2]   ——. On the linearity of testing planarity of graphs, to be published in *Annals of Chinese Math.*

[R1]   Rosenthiel, P. Preuve algebrique du critere de planarite de Wu-Liu. *Annals of Discrete Math.*, 1980, 9: 67-78.

[SB1]   Seshu, S. & Balabanian, N.. *Linear Network Analysis*. Wiley, New York, 1959.

[TU1]   Tutte, W. T.. Toward a theory of crossing numbers. *J. Comb. Theory*, 1970, 8: 45-53.

[WU1]   Wu Wen-Tsun. *A Theory of Imbedding, Immersion, and Isotopy of Polytopes in a EuclideanSpace*. Beijing: Science Press, 1965.

[WU2]   ——. Amathematical problem in the design of integrated circuits. *Mathematics in Practice and Theory*, 1973, 1: 20-40(in Chinese).

[WU3]   ——. Planar imbedding of linear graphs. *Kexue Tongbao*, 1974, 2: 226-228 (in Chinese).

[WU4]   ——. Layout problem in printed circuits and integrated circuits. Appendix to Chinese Edition of [WU1], 1978: 213—261.